Schrödinger's CAT

A Theorist's Notes on Causal Architecture

David Alfyorov

2026

Schrödinger's CAT
A Theorist's Notes on Causal Architecture

The formal SCT/CAT results presented in this book were developed jointly with Igor Shnyukov.

First edition, 2026.
Independently published.

ISBN 978-1-80645-162-3

Cover design by the author.

Set in Spectral (Production Type, OFL) and Latin Modern Math.

Author ORCID profile:
orcid.org/0009-0003-6027-7837

Author contact:
davidich.alfyorov@gmail.com

* * *

Dedicated to those who are curious but afraid to act on their passion for knowledge,

and to my friend Ruslan Ibragimov, who supported me in trying myself in a scientific environment.

* * *

What I cannot create,
I do not understand.

— Richard Feynman

Contents

Preface

I am an independent researcher; that is the part of this preface I am willing to be specific about. The rest is detail, and the detail is what this book is.

What follows is the long way round.

* * *

The pull toward the abstract

I was the kind of child for whom science fiction was insufficient because it didn't go far enough. I wanted what was actually there at the edge of physics, even if I couldn't yet read the equations. The first trapdoor through which this addiction entered my life was *Freelancer*, Digital Anvil's mostly-forgotten space simulator from 2003. It is not the game you reach for if you want to be taken seriously about space. It has a small story, an arcade flight model, mining and trading loops a child can grasp on a Tuesday evening. But its world had *names* (systems, factions, a galaxy that was clearly a piece of a larger thing), and it taught me something almost no book had succeeded in teaching me up to that age: that the unknown is not just out there. It is organised. It has structure. You can map it.

That, give or take, is what I have been trying to do ever since.

A cousin handed me *Half-Life* soon afterward. The protagonist is a theoretical physicist, MIT PhD, working on materials science at a federal research facility in the New Mexico desert. I did not yet know what theoretical physics was. I knew only that this man wore an HEV suit, ran very fast, and was the kind of person whose job it was to understand what was actually going on. That, I thought, was a job worth having.

* * *

The Wikipedia years

What followed were long nights at the family computer, reading astrophysics articles on Russian-language Wikipedia by myself, in the dark, until I could no longer keep my eyes open. Supernovae. Neutron stars. The CMB. The redshift relations. Dark matter, dark energy, the cosmological-constant problem stated naïvely as "why isn't the universe a hundred and twenty orders of magnitude smaller than it actually is." The articles were not always good. Wikipedia in those years was not what it is now. But they were doors, and the doors all opened. I read them the way other children read fantasy novels.

When physics finally appeared on the school timetable, I was already fluent in its surface. I read ahead. I asked my teachers things I shouldn't have been able to ask at that age. The class was easy.

And then it stopped being easy, in the way that things stop being easy when you are seventeen and the world has narrowed to a corridor. I will keep this brief: the last two years of school were eaten by depression, and the grades I left with were not what either I or my teachers had expected. Whatever else I want to say about that period belongs to a different book.

At graduation, my physics teacher, the one who had watched me ride the curve down, took me by the shoulder and asked for my word that I would still get a physics education. Whatever it took, whenever it came. I gave it.

* * *

The wilderness years

What followed were years that were neither education nor research nor anything you could put on a CV. I spent them, mostly, reading. In 2020 I stumbled onto the *Fermilab* YouTube channel, and Don Lincoln's lectures became the spine of my evenings. There is a particular generosity to the way Fermilab explains particle physics on the internet: they take you seriously, they don't gut the math, they walk slowly enough that you can keep up. By the end of 2021 I had a usable picture of the Standard Model, of the running of the couplings, of why the gauge sector is what it is, of why gravity refuses to be quantised the way the other interactions did. I had pictures. I had intuition. I had no way to do anything with them.

That last sentence is the one I need to dwell on for a moment, because it is the actual root of this book.

I could not code. I had never written a useful program in my life. Not Python, not Mathematica, not even a serviceable spreadsheet macro. Theoretical physics in the mid-2020s is not pencil-and-paper; it is computation, symbolic algebra, lattice construction, multi-loop sector decomposition, MCMC fits to public datasets, runs that take minutes and runs that take days. A theorist who cannot code, in those years, is a tourist holding a guidebook in a country whose roads he cannot drive. I had the maps in three languages and no way to drive.

* * *

The unlocking

Then came 2025. I tried, for the first time, to use a large language model as a research assistant. Not for chat, not for homework, but for the actual work: writing the script, debugging the integration, running the verification, reading the literature with me, arguing with me when my reasoning was loose. What had been a closed door for ten years opened on the first push. By the summer of that year I was running computations I would never have finished alone, on hardware I knew how to drive only because I had something patient enough to teach me how to drive it. The wilderness years were over, and I had not noticed they had ended until I looked back.

Several things I tried in late 2025 did not work. I will not name them; they are someone else's research questions now, not mine. But by early 2026, after enough false starts, I made a decision that on paper sounds preposterous. I would build a theory of fundamental physics. Not toy models, not phenomenology of someone else's framework, but a theory at the scale of string theory or loop quantum gravity, with its own postulates, its own mechanism, its own falsification semantic. I was not going to pretend it was anything else.

* * *

From SCT to CAT

The first form was *Spectral Causal Theory*. SCT, as I came to call it, sat squarely in the gravitational sector: a one-loop spectral effective action built on a master function $\varphi(D^2/\Lambda^2)$, parameter-free coefficients $\alpha_C = 13/120$ and $\alpha_R(\xi) = 2(\xi - 1/6)^2$ derived directly from the Standard Model field content, a no-scalaron theorem locking $\Pi_s(z, \xi) > 1$ at every Euclidean momentum and every value of the Higgs non-minimal coupling, a millielectronvolt-scale cutoff bounded from below by the LIGO gravitational-wave catalogue. It was a real theory. It also turned out, after a year of work, to be *insufficient*. There were too many things (flavour mixing, the modular structure of the cosmological constant, the value of the electroweak scale itself) that the strict $\mathcal{S}$-layer of SCT simply did not see.

Causal Architecture Theory was what came after. CAT is not a replacement for SCT; it is the parent. SCT is the strict gravitational $\mathcal{S}$-layer of CAT, and the rest (the branch package $\mathcal{B}$, with its Morgenstern algebra $\mathcal{M} = \mathbb{C} \oplus \mathbb{C}$, its cyclotomic modular field $\mathbb{Q}(\zeta_{48})$ at the period $\tau_\star = i\sqrt{2}$, and its 27-to-1 parameter census) is what I had to build to close the questions SCT couldn't reach. The book in your hands is about both, and about what is still open in the program (the $\mathcal{P}$-layer), and about what failed along the way, more candidly than most physics books permit themselves to be.

* * *

A word on the title

Schrödinger's CAT is not a pun I am proud of. It is a pun I could not avoid. The acronym was already there. The resonance with the thought experiment was unavoidable. And the spirit of the book (a theory that, in one set of measurements, sits clearly alive in the box, and in another set of measurements, sits clearly dead) is the exact position from which I want to write.

Not all of CAT is alive. Not all of it is dead. A non-trivial fraction is genuinely in superposition, and the only honest thing to do is to open the box and report what was inside, including the parts that smelt of formaldehyde.

* * *

What this book is, and what it isn't

What this book does *not* do is claim that quantum gravity has been solved. It does not claim to be the last word on anything.

It does claim that, within the canonical branch of CAT, twenty-seven free parameters of the Standard Model and cosmology collapse to one, and that this collapse is testable. It claims that the gravitational sector at one loop is parameter-free in a sense that can be checked against gravitational-wave bounds. It claims that several once-promising routes (measure-based dark energy in $C_{\mathrm{diag,HS}}$, linearised singularity resolution at the black-hole centre, asymptotic safety with CAT boundary conditions for the electroweak scale) are dead, and have stayed dead under attempts to revive them. The book is for the reader who wants to know which.

The chapters alternate between the formal physics (derivations, status taxonomy, theorem statements with their conditions) and *naked notes*: first-person, dated, honest interludes about what I thought a year ago, what made me change my mind, what scared me, what I still don't know how to close. The naked notes are the spine of the book, in the sense that they are the part that explains why the formal physics has the shape it has.

This is not the book the field will eventually write about CAT. That book, if it gets written at all, will be written by someone else, after the dust settles. This is the book *I* could write, in 2026, with the door behind me still open and the work still in progress.

David Alfyorov

Prologue: What CAT does and what it doesn't

The shape of the dissatisfaction

I came to this work having spent fifteen years reading, on and off, the two great cathedrals of theoretical physics that dominated the field through my entire intellectual life: string theory and loop quantum gravity. Let me say at the start of this book what I think of them, because everything that follows is a reaction to that thinking.

I admire string theory the way one admires a system of mathematics so deeply worked out that it has acquired a life of its own. Its internal consistency is real. Its mathematical depth is real. Its contributions to pure mathematics, and to the deep abstract discovery that quantum theories which look very different on the surface can secretly be the same theory in disguise, are not a marketing fiction. They are part of the permanent record of late twentieth and early twenty-first century theoretical physics. I do not believe what some of its critics believe.

But I also do not believe what some of its proponents believe. In nearly half a century, string theory has not produced a single falsifiable prediction at any energy scale that any present or planned experiment will reach. Its landscape (the menu of universes the theory says are mathematically allowed) contains something on the order of 10^{500} entries, and there is no known principle that picks out ours. The choices that distinguish one of those universes from another are, for the most part, not even principled distinctions; they are choices made for the sake of making the math tractable, with no inner mechanism demanding them. A theory whose answer to which universe is ours has to be "the one in which observers like us happen to exist" is not, in the strict sense in which I want to use that word, a physical theory of the world we live in. It is a theory of every possible world, with a footnote saying that ours is one of them.

Loop quantum gravity is the other cathedral. It is smaller, less crowded, and (I think) more honest about what it is doing. It begins with a clear move (try to apply the standard quantum mechanical recipe to Einstein's gravity directly, without picking a fixed background of space and time first), and it pursues that move with discipline. I have a great deal of sympathy for what it is trying to be. But the question of how the smooth spacetime we actually see in everyday life emerges from this granular picture is, after more than thirty years, still not closed in a way that satisfies me. The way the theory talks to ordinary matter is not where one would want it to be. And the experimental handles are, again, indirect: predicted graininess scales for spacetime that lie far below anything we can currently measure, predicted distortions in how light of different colours travels through space whose magnitudes sit so far below current bounds that "not yet ruled out" is doing more work than it should.

I do not say either of these things to be polemical. I say them because I read both literatures with attention, in the years when I had nothing else to do but read, and I came out of those years with a particular kind of hunger. I wanted a theory of fundamental physics that could be *killed*. Not in principle, not at scales sixteen orders of magnitude above anything the LHC will ever reach, but killed by data we already have, or data we will have within a decade, or data the next gravitational-wave catalogue will hand us without asking.

* * *

The trade I was willing to make

CAT is what came out of that hunger, and CAT is built on a trade that is the whole reason the book exists.

I gave up, on day one, the ambition of writing down a complete theory of quantum gravity. That door is still closed in CAT. The biggest things any candidate theory of quantum gravity is supposed to deliver, the questions every serious framework on the market points at as its eventual prize (whether the math stays sensible at arbitrarily small distances, whether cause and effect hold up everywhere, whether probabilities still add up to one hundred per cent in every calculation, whether the singularity at the heart of a black hole gets smoothed away), are still being worked on inside CAT. Some of them I have made progress on. Most of them I have not closed. The book labels them as open and does not pretend otherwise.

What I bought with that concession was something else. The gravitational sector of CAT (the part the book calls the $\mathcal{S}$-layer) has a remarkably tight structure. The dominant correction to Einstein's gravity at the simplest quantum level is fixed by a single number, $\alpha_C = 13/120$, which drops out of counting the fields in the Standard Model and is not a parameter I can tune. A theorem keeps a particular slice of the math positive everywhere it has to be, which means no fifth force can sneak into the lab through a back door, no matter how the Higgs particle is coupled to gravity. The only free energy scale in the theory, called Λ, is already bounded from below by the gravitational-wave detectors at LIGO, Virgo and KAGRA, which have already ruled out any value smaller than about $8.50\,\mathrm{meV}$. And in the canonical version of the branch package, twenty-seven free numbers in the Standard Model and cosmology collapse to a single one, and that collapse is a check on the theory, not a fit.

This is not a theory of everything. It is a theory of fewer things, done in a way that the next round of experiments can disagree with. Every claim in this book sits on a falsification semantic that says, in advance, what would kill it. The atlas of those kill-switches is in the back. The book has been written so that, if you want to, you can turn to that atlas first.

That is the trade. The chapters that follow are an account of what I bought and what I gave up to buy it.

* * *

The theory-of-everything market

Let me widen the lens for a paragraph or two. The dissatisfaction is not only with the two cathedrals. It is with the modern theory-of-everything scene as a whole, and it is worth saying out loud what is wrong with it.

In the years I was reading, the field of fundamental theory did something strange. It did not converge. It bloomed. Asymptotic safety, causal dynamical triangulations, group field theory, spin foam models, twistor theory, emergent gravity, E_8 unification, every flavour of brane-world phenomenology and extra-dimensional construction, every flavour of holographic speculation, plus a long tail of bespoke frameworks each with its own postulates, its own moduli space, its own internal vocabulary. A lot of this is serious work by serious people, and I do not want to belittle any of it. But taken together they form a market, and the market has the characteristics of a market: more producers than consumers, more proposals than tests, more arXiv preprints per week than human attention can absorb in a year.

The deeper irritation is structural, and it is the same irritation in every case. Each of these programmes, taken individually, is consistent with current data, because current data is sparse exactly where the theories disagree. Each of them, taken individually, predicts that some delicate effect will appear at an energy or sensitivity scale we cannot yet reach. Each of them, taken individually, has a tunable parameter or a moduli choice that lets its predictions slide back into "not yet ruled out" the moment the data threatens. The result, viewed from outside the sociology, is that in 2026 we have on the order of a hundred mutually incompatible candidate theories of fundamental physics, and exactly zero of them are falsified. That is not what a healthy field looks like. That is what a stalled field looks like.

There is a separate sub-genre worth naming, because it deserves its own paragraph. In the last decade, a class of "computational" or "information-theoretic" theories of everything has acquired a public profile that wildly exceeds its technical content. Hypergraph rewriting systems sold as the substrate of physics. Cellular-automaton universes pitched as final theories. "The universe is a quantum computer" frameworks with no quantitative output. Constructor theory in its more inflated public-facing forms. The various it-from-qubit programmes when they overreach. Some of these contain genuinely interesting mathematics, and a few have serious researchers behind the technical content. None of them, at the time of writing, has produced a single quantitative prediction an experimentalist can take into a lab and check. I do not have the patience for theories whose primary medium of dissemination is the popular interview. Physics that is confident enough to talk to journalists before it has produced a falsifiable prediction is doing something other than physics.

CAT is not a reaction against any one of these programmes. It is a reaction against the shape of the whole landscape. The shape is this: too many candidates, too few tests, too long a horizon between proposal and verdict. I wanted, for once, the verdict horizon to be shorter than my own career. That is the design principle this book is written around.

* * *

CAT and SCT, and a thing I have to confess about gravity

Two acronyms run through this book, and the relationship between them is the kind of thing I would rather sort out on the first page than have you wonder about for two hundred. Before I get to the acronyms, though, I owe you a confession that frames the whole project.

I have never loved the idea that gravity has to be quantized.

To most physicists, this is the central unsolved problem of the field. Gravity, alone among the four known fundamental forces, has stubbornly refused to be made compatible with the rules of quantum mechanics that work so beautifully for everything else. There is a hundred-year-old hole in the textbooks waiting for somebody to fill it. There is an entire industry trying. And on a purely aesthetic level, I never wanted to be one of those somebodies.

The reason is, I think, easy enough to feel. Gravity, the way Einstein left it, is geometry. It is the shape of spacetime itself, smooth and continuous, the same way the surface of a trampoline is smooth even when something heavy is sitting on it and bending it down. Quantum mechanics, on the other hand, is the physics of things that come in pieces. Light comes in photons. Matter comes in electrons and quarks. Energy comes in lumps you can count. To take something as smooth and stately as the geometry of the universe and chop it into countable pieces feels, on the face of it, like a violation of the spirit of what gravity actually is.

But here is the trouble. Nature does not seem to care what I find aesthetic. Every single time, in the last hundred years, that we have looked closely at something we thought was smooth, we have found it grainy underneath. Light, looked at closely: grainy. Matter, looked at closely: grainy. The vacuum itself, looked at closely enough with sensitive enough instruments: grainy. The pattern, repeated for over a century now, is that the universe loves discreteness. Smooth is what things look like from far away. Lumpy is what they actually are.

So I made a tradeoff. I do not love the idea of quantizing gravity. But the universe seems to want me to, and the universe gets a vote. The version of that compromise I could live with is what I built first, in 2025, and I named it SCT.

Spectral Causal Theory, SCT, is the gravitational core of this book. It is not a complete quantum theory of gravity. It does not pretend to be one. What it is, instead, is a careful and conservative description of the small corrections that quantum mechanics adds to Einstein's smooth picture, in the regime where those corrections first start to matter, in a form that can be checked against experiments we are already running today.

The shape of SCT is tight enough to describe in one paragraph. There is one master function (a particular curve, computed once, that controls how gravity behaves at short distances). There is a parameter-free number, $\alpha_C = 13/120$, that drops straight out of counting up the fields of the Standard Model and is not something I can tune. There is a theorem locking out the worst kind of unwanted side effects, the ones that would show up in the lab as a fifth force pulling on objects in ways gravity is not supposed to. And there is a hard lower bound on the only free energy scale in the theory, Λ, that comes straight out of the gravitational-wave detectors at LIGO, Virgo and KAGRA: any value below $\Lambda > 8.50\,\mathrm{meV}$ is already ruled out by data on the shelf. I was happy with it when I finished. I am still happy with it. It is what shows up in this book as the $\mathcal{S}$-layer.

Then I tried to push SCT outward, and that is where I lost. The Standard Model has flavour, the property that distinguishes a muon from an electron and a strange quark from a down quark, and SCT did not see flavour. The cosmological constant has a value (the small but nonzero amount of energy that fills empty space and drives the universe to expand faster), and SCT had no way to predict that value. The electroweak scale, the energy at which the weak nuclear force changes its character, exists, and SCT had no mechanism for it. I spent a couple of months in early 2026 trying to make SCT cover those things, and I lost cleanly. SCT was the wrong-shaped tool for those questions. So I built a bigger one around it.

* * *

Who doesn't like CATs? We certainly do.

Causal Architecture Theory, CAT, is what came out of that. SCT sits inside CAT as the gravitational layer. Around it sits the branch package, an algebraic and modular structure that turns the flavour question and the cosmological-constant question into theorems, conditional on a single choice. Around that sits the open program, the things CAT cannot yet do, which I will not pretend it can.

When I say "the theory" from this point on, I mean CAT, with SCT inside it. That is the order I built them in, and that is the order this book reads in.

* * *

The S/B/P stratification

CAT has three layers. They have different epistemic statuses, and the book treats each of them differently. Pretending that all three sit at the same level of certainty would be a kind of dishonesty I do not want to commit on page one.

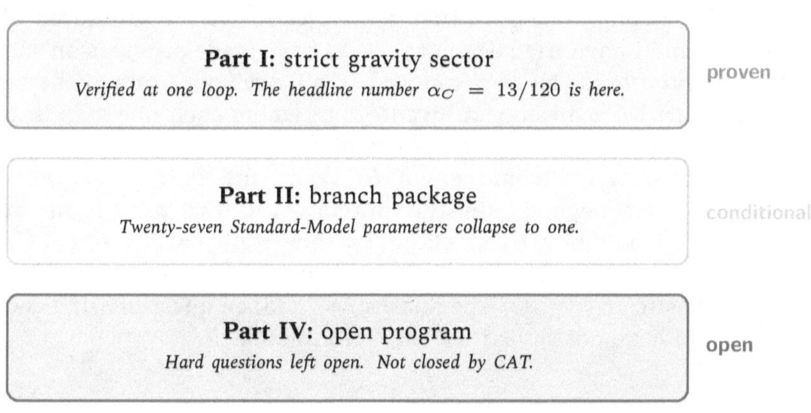

Three layers of confidence: from theorem-grade rock to open frontier.

The first layer is the gravitational core. The $\mathcal{S}$-layer. Everything in it has been derived, verified, and stress-tested to the level a journal referee would expect: the one-loop coefficient, the no-scalaron theorem, the cutoff bound from gravitational waves, the limits back to general relativity and the Standard Model and quantum mechanics. If you read only Part I, you will read the strongest part of the book. It is the part I am most willing to defend in a hostile seminar room.

The second layer is the branch package. The $\mathcal{B}$-layer. Here the statements are still theorems, but they are conditional: they hold given a choice of which branch we are on, and the canonical branch is the one this book uses. On that branch, the flavour structure of the Standard Model assembles cleanly out of a small algebraic skeleton. The modular layer fixes the mixing angles to numbers I did not choose. The cosmological constant comes out as an arithmetic expression. Most strikingly, the parameter count drops from twenty-seven free numbers to one. Part II is about this layer. It is the most surprising part of the book, and it lives or dies on the branch choice.

The third layer is the open program. The $\mathcal{P}$-layer. This is where the things any serious theory of quantum gravity is expected to do are still being worked on. Does the math stay sensible at arbitrarily small distances (what physicists call full UV-finiteness)? Does the structure of cause-and-effect hold up everywhere across the manifold (global Lorentzian closure)? Do probabilities still add up to exactly one hundred per cent no matter how baroque the calculation gets (all-orders unitarity)? Does the singularity at the centre of a black hole stop being a singularity (full nonlinear singularity resolution)? Can the energy at which the weak nuclear force changes its character be computed from first principles instead of measured (the v_{EW} derivation)? I have tried each of these. I have made progress on some. I have failed on others. None are closed. CAT does not pretend they are closed, and Part IV is an honest inventory of where each one stands and where I think it might or might not go from here.

So the three layers come down to this: the part you can rely on, the part that depends on the branch, and the part I am still working on. Every nontrivial claim in this book carries one of six status tags ([PROVEN] proven, [VERIFIED] verified, [CONDITIONAL] conditional, [HEURISTIC] heuristic, [SPECULATIVE] speculative, [OPEN] open question). If a claim is not tagged, I have not yet earned the right to make it.

* * *

The naked-notes promise

One last thing before the formal chapters start.

Inside almost every chapter you will find italic interludes set off by thin rules above and below, prefixed **Naked notes.** These are the personal voice of the project. They are dated. They are in the first person. They report failures, abandoned routes, things I believed a year ago and no longer believe, intermediate states of mind that turned out to matter to the eventual physics in ways that, if I cleaned them out, would make the formal chapter shallower than it ought to be.

Most physics books polish the failures out, and the reasoning is respectable: a textbook is a finished object, and the dead branches of the tree do not belong in a finished object. I have a different intuition for this book. The shape of a fundamental theory, in its first written-down form, is not understandable without the abandoned shape that came before it. The naked notes are an attempt to keep that visible instead of throwing it away.

A small discipline. Footnotes are reserved for citations and short technical asides only. Anything that would otherwise have been a personal comment, a memory, a confession, or a methodological remark goes into a naked-notes block. The parts of the book that are claims about physics and the parts that are claims about how the physics got made are not in the same paragraph, on purpose.

If you are reading this for the formal results and find the naked notes intrusive, skip them. They are visually marked and easy to skim past, and the chapters stand without them. But I would suggest reading at least the first two or three before you decide. They were written for the reader who wants to know not just what the theory says, but why it has the shape it has.

How to read this book

This is a hybrid book. It carries the skeleton of a serious work of theoretical physics, with eight verification layers behind every VERIFIED result, and the prose of a popular-science book, in which the math is explained rather than demanded. The two registers do not interfere; they are layered the way a good film score is layered behind dialogue. You can listen to either, or both.

What follows is a short menu of the routes through the book, so the reader who is short on time, or who is here for one specific question, does not have to read the whole thing in order to get their money's worth.

The five-minute reading. Read this page, the preface, and the canonical-numbers appendix on page 234. That is everything the theory gets right, with status tags and chapter references, in three pages. If you want one number from this book, that is the page that has all of them.

The thirty-minute reading. Read the prologue (page xi), Part I, and the epilogue (page 216). The prologue tells you what kind of theory this is. Part I is the strongest sector, with $\alpha_C = 13/120$ derived from first principles. The epilogue lists ten kill-switches: specific things that, if measured, would falsify the theory. One tight pass through the engine.

The full reading. The book in its intended order. Part I (gravity, theorem-grade), Part II (the branch package and the twenty-seven-to-one census, theorem-grade conditional on the canonical branch), Part III (cosmology), Part IV (the open frontiers), and Part V (the method: how I know when I am right and how I avoid fooling myself). About two evenings, at the pace of a cosy non-fiction read.

The reader who is here for one thing. Three short paths for specialised readers.

- For *the gravity story*: Part I (Chapters 1 through 6).
- For *the flavour predictions* (PMNS, CKM, Ω_Λ): Part II (Chapters 7 through 10).
- For *the methodology and the LLM-era authorship question*: Part V (Chapters 20 through 23).

The naked notes. Inside almost every chapter you will find an italic, dated, first-person interlude, set off in a tinted box and prefixed by the words *Naked notes*. These are deliberate: they mark the moments where I want you to know what the author was actually thinking, not just what the published version says. If you treat them as optional asides you will still get the physics. If you read only the naked notes, in order, you will get a different book: a methodological diary of how the work was done. The pop-science prose and the diary are co-equal, and the book asks you to take both seriously.

PART I	**Gravity (theorem-grade).** The form factor, the no-scalaron theorem, the Λ-bound, and the discrete-causal universe.	*5-minute read:* canonical numbers
PART II	**Branch package (conditional).** The Morgenstern algebra, the modular point, and the twenty-seven-to-one census.	*30-minute read:* prologue + Part I
PART III	**Cosmology.** Dark energy, dark matter, inflation, black-hole entropy, and the modified Friedmann equation.	*full read:* five parts in order
PART IV	**Open frontiers.** UV-finiteness, singularity resolution, and the three parallel routes to the electroweak scale.	
PART V	**Method.** How the work was actually done. The eight-layer verification pipeline, fifteen ways to fool yourself, three minds arguing every result, and honest theory in the LLM era.	

A map of the book: five parts, three suggested reading paths.

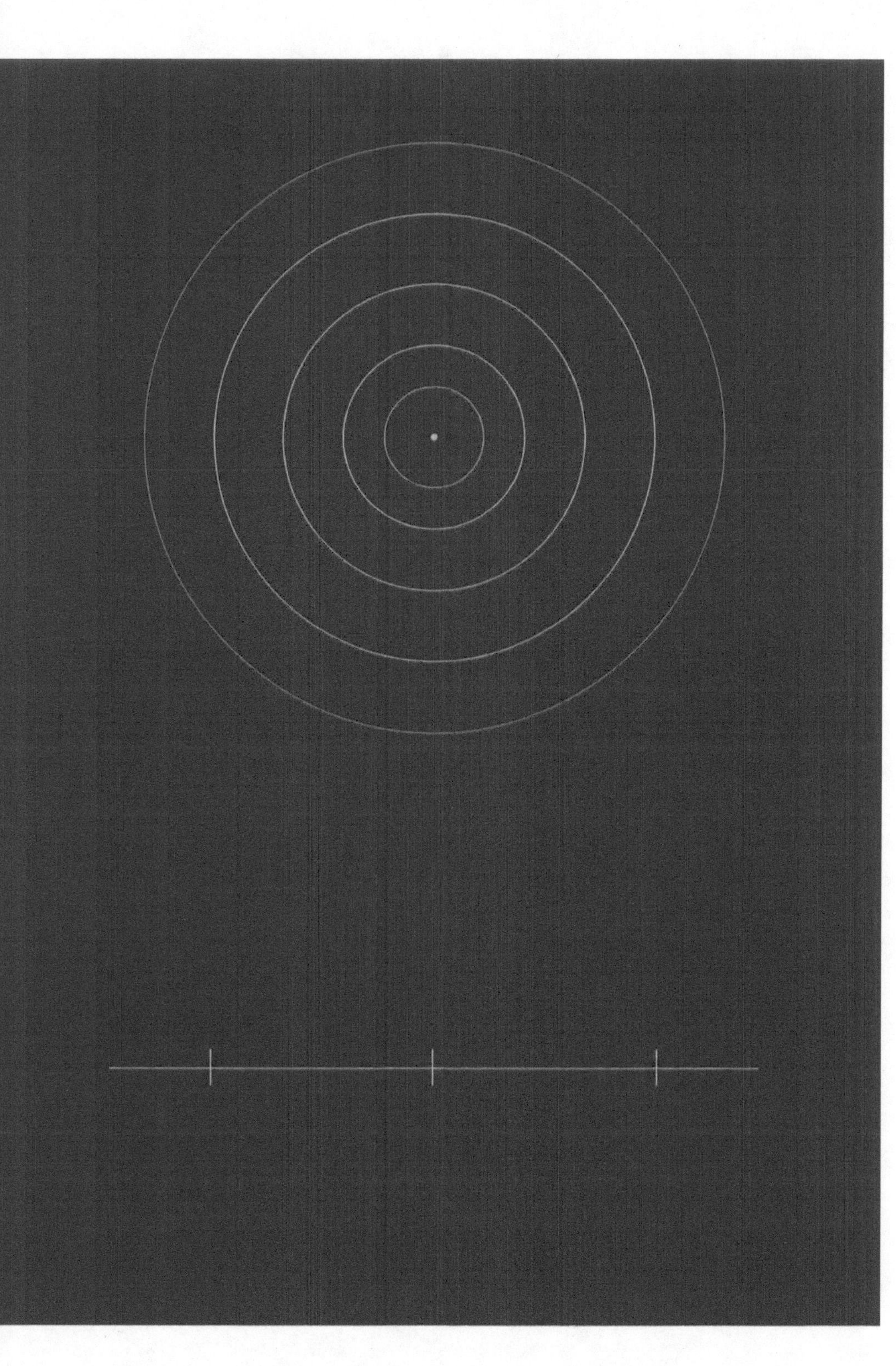

PART I

SCT: Gravity ($\mathcal{S}$-layer)

Six chapters about gravity, told roughly in the order I figured them out. A single curve, computed once, that controls how the quantum world modifies Einstein's smooth picture at very short distances. A perfect-square structure that rules out a fifth force on the next two pages. A small rational number, 13 over 120, that falls cleanly out of counting up the kinds of particle in our universe. A bound from the gravitational-wave detectors that puts the only free dial of the theory in the millielectronvolt range, with the next decade of measurements lined up to either confirm or kill it. A chapter on how to recover all of this from inside a discrete network of events, with a new observable that detects the ripples of gravity where the standard tools cannot reach. And a final, optional chapter on the handedness hidden inside the heat-trace calculation that underlies it all. If you read only one part of this book, this is the part.

CHAPTER 1

The shape of gravity, up close

The miracle of the appropriateness of the language of mathematics for the formulation of the laws of physics is a wonderful gift which we neither understand nor deserve.

— EUGENE WIGNER, 1960

Two numbers are going to come out of this chapter, and the rest of Part I hangs on them. The first is a single parameter-free coefficient,

$$\alpha_C \;=\; \tfrac{13}{120},$$

that fixes the most important quantum correction to Einstein's gravity in the regime where I can actually compute things. The second is a coefficient $\alpha_R(\xi)$ that depends on a single dial ξ, the way the Higgs particle is coupled to gravity, and has the shape

$$\alpha_R(\xi) \;=\; 2\big(\xi - \tfrac{1}{6}\big)^2 .$$

That second formula is the seed of the no-scalaron theorem in Chapter 2 and the gravitational-wave bound on the cutoff Λ that closes Part I. Both numbers fall out of nothing more than the field content of the Standard Model. They are not parameters I picked. They are not parameters anyone gets to pick.

The shape of gravity, up close

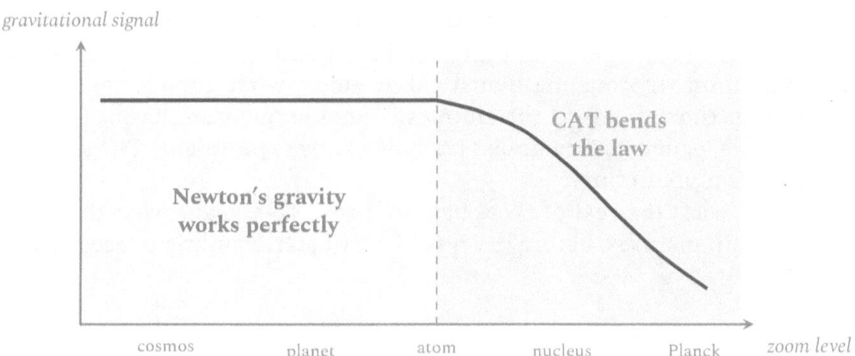

Where gravity looks normal, and where it doesn't.

* * *

1.1 Where Einstein leaves off

At the scales where humans live — a dropped apple, the moon's orbit, the slow inspiral of the Hulse–Taylor binary pulsar PSR B1913+16 over decades — gravity is what Einstein wrote down a hundred and ten years ago. Spacetime has a shape. Mass and energy bend the shape. Things follow the bent shape from inside it. The math works to ten significant digits everywhere we have ever checked.

But Einstein's theory, like every theory in physics, has a regime of validity, and the regime gets narrower as the scales get shorter. At ten centimetres, perfectly fine. At ten micrometres, fine. At ten nanometres, still fine. By the time you get down to what physicists loosely call "short distances" (somewhere between the size of an atom and the energy frontier of our particle accelerators, or, equivalently, between everyday energies and the energies a collider like the LHC reaches), the smooth picture has to start cracking, because quantum mechanics is real and quantum mechanics insists on lumps.

What does the cracking look like? That is the question this chapter is about, and that is the question CAT is built to answer in a particular, restricted, but verifiable way.

Most candidate theories of fundamental physics give an answer of the form "We do not know, but here is a set of free parameters that will let us fit whatever turns up." The strict gravitational sector of CAT does not work that way. It says: here is a single curve. The curve has one free energy scale on it. The shape of the curve is not a parameter. The shape of the curve is what falls out of summing up the field content of the Standard Model and asking, on rigorous mathematical grounds, what gravity does in the presence of those fields at the simplest possible quantum level. The free energy scale, called Λ, is bounded from below by experiment. The shape of the curve is a prediction.

That is what the rest of this chapter is for. To explain what the curve actually is, in pictures, before the rest of Part I starts pulling consequences out of it.

* * *

1.2 Form factors, by analogy

Drop a stone into a still pond. Ripples spread from the splash. How fast they move, how their amplitude falls off with distance, what happens when two of them meet — in a perfectly clean pond, this is high-school physics: a single-frequency wave with a known dispersion relation.

Now make the pond more interesting. Put leaves on the surface. Add swimming fish. Stir up a current. The ripples are still ripples, but their shape now depends on what is in the pond. The high-school formula is no longer enough on its own. You need a correction that knows about the leaves, the fish, the current. Physicists call this kind of correction a *form factor*.

Gravity, in Einstein's original picture, is the pond when nothing else is in it. Pure geometry, pure ripples, pure mathematics. The moment you remember that the rest of physics is also there (the electron, the photon, the quark, the Higgs, the entire matter content of the universe sitting inside spacetime), the gravitational ripples acquire a form factor. The form factor tells you how the matter inside changes the way the ripples look at short scales.

CAT computes this form factor. It does not propose it; it does not parametrise it; it does not leave it as a fitting parameter. It computes it once, from the field content of the Standard Model, in the simplest quantum regime where the question makes sense.

* * *

1.3 The master function

The output of that computation is a single curve, the same curve everywhere in the theory, which I will call φ (phi). It is the spine of the gravitational sector of CAT. Everything else in this part of the book hangs off it. Its formula is

$$\varphi(x) = e^{-x/4}\sqrt{\frac{\pi}{x}}\,\mathrm{erfi}\left(\frac{\sqrt{x}}{2}\right), \tag{1.1}$$

where erfi is a standard mathematical function (the imaginary error function, found in any handbook of special functions) and x is a dimensionless ratio: the energy scale you are probing, divided by a single energy Λ that is the only free dial in the gravitational sector of the theory.

You do not need to read the formula to understand what it says. What it says is this.

At long distances, φ goes to one. This is the limit $x \to 0$, where you are probing scales much larger than the inverse of Λ. In this regime the form factor turns off and gravity reduces, exactly, to what Einstein wrote down. VERIFIED The Taylor expansion at $x = 0$ is $\varphi(0) = 1$, $\varphi'(0) = -1/6$, smooth and analytic, with no surprises. This is why Einstein's theory works so well in the lab and in the solar system: the cloud-correction the master function carries is invisible at the scales where humans live.

At short distances, φ falls. As x grows, the master function decreases, and the form factor turns on. The quantum cloud of Standard-Model matter starts to affect the way gravity propagates. The change is not a free knob anyone can turn. It is dictated by the formula above.

The transition happens at a single energy scale, Λ. This is the only free parameter the gravitational sector of CAT contains. Everything else gets computed once and for all. Λ is bounded from below by experiment: the gravitational-wave catalogue from the LIGO, Virgo and KAGRA collaborations already gives us VERIFIED $\Lambda > 8.50\,\mathrm{meV}$, a bound several orders of magnitude away from any current laboratory or particle physics measurement. Chapter 4 is about that bound, what it means in everyday terms, and why it is a real prediction the theory was forced to make.

That is the whole content of the master function in three paragraphs. One curve, one free scale, two end behaviours. Out of this single object, the theory builds the form factor; and out of the form factor falls every result in the strict gravitational sector of the book.

* * *

1.4 Two coefficients

The form factor is not a single number; it is two numbers, because the gravitational field has two natural pieces. Knowing this small piece of structure makes the rest of the chapter make sense.

The Weyl piece, and the number 13/120

The first piece of the gravitational field is what physicists call the *Weyl* part. In picture terms, it is the part of gravity that lives *between* massive bodies. It is the geometry of the ripples themselves, the part of the field that propagates out into empty space far from any matter. It is what gravitational waves are made of. It is the part of gravity that LIGO sees.

The coefficient that fixes how this piece of gravity is modified at short distances, summed over the whole field content of the Standard Model, comes out of the central computation of the strict gravitational sector. The answer is

$$\text{VERIFIED}\ \alpha_C = \frac{13}{120}. \tag{1.2}$$

A rational number. Three small primes in the numerator, two small primes in the denominator. Not adjustable. Not scheme-dependent. Not the result of a fit to any data. It comes out of counting how many scalars, fermions and gauge bosons live in the Standard Model spectrum, weighting each one with a known coefficient that one can look up in any handbook of one-loop gravity, and adding everything up. Chapter 3 unpacks where the thirteen and the hundred-and-twenty actually come from. For now, all you need to know is that the answer is parameter-free and that this is unusual.

The Ricci piece, and the perfect square

The second piece of the gravitational field is what physicists call the *Ricci-scalar* part, written R^2. In picture terms, it is the part of gravity that talks directly to whatever matter is sitting in the spacetime: the piece that couples to ordinary stress–energy, the piece that, if it propagated as an independent particle, would show up in a laboratory as a fifth force pulling on objects in ways that ordinary gravity is not supposed to.

The coefficient that fixes this second piece is $\alpha_R(\xi)$, and unlike α_C, it depends on one number. That number is the parameter ξ that controls how strongly the Higgs particle is coupled to gravity. The result, derived in the same one-loop computation that fixed α_C, has a particular shape:

$$\text{VERIFIED}\ \alpha_R(\xi) = 2\left(\xi - \tfrac{1}{6}\right)^2. \tag{1.3}$$

It is a perfect square. It is non-negative everywhere. At one specific value of the Higgs–gravity coupling, namely $\xi = 1/6$ (called *conformal coupling,* because at that one value gravity coupled to the Higgs is invariant under local rescalings of distance — a deep symmetry that pulls the coefficient down to nothing), the coefficient α_R vanishes exactly. Off conformal coupling it is small but not zero, growing as the square of how far you sit from $\xi = 1/6$.

That perfect-square structure is the fingerprint of a hidden symmetry. Most of the particles in the Standard Model, the electrons and quarks of everyday matter, the photons of light, the gluons and W and Z bosons of the nuclear forces, all preserve a certain symmetry called conformal invariance at the simplest quantum level in four dimensions. Conformal invariance means, roughly, that whether you look at the math from a millimetre or a micrometre away, the equations look the same up to an overall scaling. Particles with that symmetry contribute zero to α_R. The only matter field in our universe that breaks conformal invariance is the Higgs, and it breaks it by exactly the amount $(\xi - 1/6)$. Squared, because the contribution is the modulus squared of an interference and we are summing positive things. That is the entire derivation of the formula, in five lines of English.

This perfect-square structure is the seed of every single result in the rest of Part I. It is why no scalar particle propagates in the gravitational sector of CAT (Chapter 2). It is why the gravitational-wave bound on Λ comes out as clean as it does (Chapter 4). The whole gravitational story rests on the fact that one perfect square and one rational number are enough to determine everything.

* * *

1.5 Why you can rely on these numbers

A coefficient like $\alpha_C = 13/120$, written down in a pop-science book by an independent researcher with no institutional letterhead, is the kind of thing you should be suspicious of by default. I would be suspicious of it. So would any practising physicist who knows how easy it is to convince oneself, with sloppy work, that some number one wanted to be true is true. Before the rest of Part I starts pulling consequences out of these numbers, here is what was done to make them trustworthy.

Internally I have been calling the procedure the *eight-layer verification pipeline.* The full story sits in Part V; what follows is the short version.

- **Layer 1, analytic.** Dimensions, limits, sign conventions, pole cancellations: the math works out to the rules every physics textbook teaches.
- **Layer 2, numerical.** The same expression, evaluated by an independent computer-algebra path to over a hundred digits of precision, agrees with the analytic result.
- **Layer 2.5, property fuzzing.** A thousand random configurations are thrown at each formula, checking for violations of properties (positivity, monotonicity, consistency under parameter changes) that the formula has to obey on general grounds.
- **Layer 3, literature.** The intermediate quantities the calculation passes through also appear in published papers by other authors, working on adjacent problems. The numbers I get and the numbers they published agree, when both are translated into the same conventions.
- **Layer 4, dual derivation.** The same final answer is produced by a second derivation that follows a different path through the math. If a sign or a factor went wrong in the first path, the dual catches it.
- **Layer 4.5, triple computer-algebra.** Three different computer-algebra systems (each one written by a different team, each with its own bugs and quirks) produce the same intermediate symbolic expressions and numerical values.
- **Layer 5 and Layer 6, formal proof.** The key rational identities are translated into a formal proof assistant and checked by a machine that has no intuition, no taste, and no way of being talked into the wrong answer. If a logical step went wrong in a way humans miss, the proof assistant flags it.

If you read only the chapters in Part I, you are reading the part of the theory that has been put through this entire pipeline and came out clean. None of the results in this Part are conditional on a branch choice. None of them are speculative. The status tags next to each result will say VERIFIED or PROVEN, and they will mean it.

For $\alpha_C = 13/120$, the pipeline ran clean across VERIFIED the combined Standard Model computation, extensively cross-checked across all eight verification layers. For the no-scalaron theorem in Chapter 2, the verification was at the linearised level with a separate triple-redundancy audit. These checks are worth mentioning, even in a pop-science book, because they are the honest answer to the natural question, "how do I know this is not just one person being convinced of something he wanted to be convinced of."

• • •

NAKED NOTES. 2026-04-29.

The first time I computed α_C at the end of 2025, I was sitting at the kitchen table with a laptop and a cold cup of tea, and the symbolic algebra system was running through what I think was its third independent re-derivation in a row. I had told myself I would be happy with anything that fitted onto a single page of A4 paper. In my head, I was expecting a long irrational expression with logarithms, possibly a polylogarithm or two, and a coefficient that depended on the renormalisation scheme.

What came out was a fraction. Thirteen over a hundred and twenty. No logs. No polylogs. No scheme dependence. Three small primes in the numerator, two small primes in the denominator, and a structure that, when I traced it back through the per-spin contributions, was visibly dictated by the field content of the Standard Model in a way I had not put in by hand.

I sat at the kitchen table for a while after that, because if you have ever been an independent researcher with no institutional backing, and you are looking at a clean rational answer that you did not steer toward, you spend a few minutes doubting that you did the calculation correctly. I re-ran the calculation in three more ways before I let myself write the number down anywhere permanent. I re-ran it a fourth way the next morning. By the third week, I had it inside the formal-proof assistant with two independent backends agreeing. By that point, 13/120 had become as solid an object as anything I had ever computed.

I am writing this Part of the book around that number. If at some point in the next few years a careful experimentalist measures the gravitational form factor in a regime sensitive to the α_C coefficient and finds a value cleanly inconsistent with 13/120, the strict gravitational sector of CAT is dead. I have to be honest that this is the deal.

---

* * *

1.6 What's coming in the rest of Part I

Chapter 2 takes the structure $\alpha_R(\xi) = 2(\xi - 1/6)^2$ and shows that it is enough, all by itself, to rule out any real scalar particle in the gravitational sector of CAT. There is no fifth force. Not because I assumed it away, but because the structure of the form factor forbids it.

Chapter 3 unpacks $\alpha_C = 13/120$ in detail: where the thirteen comes from, where the one hundred and twenty comes from, and what would happen to the number if you replaced one of the Standard Model's fields with something else. Some replacements break the prediction. Others do not. We will see why, and what the size of the safety margin is.

Chapter 4 turns to data. The cutoff Λ that controls when the form factor turns on is the only free parameter in the gravitational sector. The current gravitational-wave catalogue gives us $\Lambda > 8.50\,\mathrm{meV}$ already. We will see what this number means in everyday terms, what it means for the existence of new physics at scales below it, and what observations would tighten the bound further as the next runs of LIGO and Virgo come online.

OPEN Part I does not cover the open program: the nonlinear extension of the no-scalaron theorem, the all-orders ultraviolet behaviour of the theory, the Lorentzian-closure problem. Those live in Part IV, which is labelled as open program for a reason.

CHAPTER 2

No-scalaron, and the absence of a fifth force

I have done a terrible thing, I have postulated a particle that cannot be detected.

— WOLFGANG PAULI, 1930

The previous chapter ended on the formula

$$\alpha_R(\xi) = 2\left(\xi - \tfrac{1}{6}\right)^2.$$

This chapter is about what that formula does. The single perfect-square structure, together with one short positivity argument, is enough to rule out the existence of any real scalar particle in the gravitational sector of CAT. There is no fifth force in this theory. Not because I assumed there isn't one, but because the math forbids it.

The result is a theorem. It is not a guess. It has been proved at the linearised level in the field equations, with the Standard Model matter content and explicit numerical robustness margins, and audited three independent times. I will state it precisely later in the chapter. The status tag attached to it is PROVEN.

* * *

2.1 What a fifth force would have looked like

Before I can explain why CAT does not have a fifth force, it is worth a paragraph on what a fifth force *is*, and why theorists have been looking for one for nearly half a century.

In the standard textbook picture, the gravitational field has only one kind of propagating particle: the graviton, a massless spin-2 quantum that carries the ripples of curvature from one point of spacetime to another. In any theory of quantum gravity that goes beyond Einstein's, however, there is a real risk that one accidentally introduces an extra particle on top of the graviton. The most common extra particle is something physicists call a *scalaron*: a single scalar quantum, no spin, just a number at every point of spacetime, that would propagate alongside the graviton and would couple, in some way, to ordinary matter.

If a scalaron existed and were light enough to propagate at laboratory or astrophysical scales, it would show up in measurements as what physicists call a fifth force: a force, distinct from gravity, electromagnetism, and the two nuclear forces, that would pull on objects in ways the four known forces do not. Decades of high-precision experiments (torsion balances, lunar laser ranging, Cassini-spacecraft tracking, atom interferometry) have looked for exactly such a fifth force and have not found one. The current bounds say: if a fifth force exists at all, it must be much weaker or much shorter-ranged than ordinary gravity at every scale we have ever probed.

A theory of quantum gravity that predicts a low-energy scalaron is therefore in trouble with the data. Any candidate theory has to either suppress its scalaron explicitly, hide it behind some mechanism, or, ideally, show from first principles that no such particle is allowed in the first place.

CAT does the third thing.

* * *

2.2 Where the perfect square comes from

The starting point is the formula at the top of this chapter. The total contribution of the Standard-Model matter to the R^2-coefficient (the part of the form factor that controls the Ricci-scalar piece, the piece that talks to ordinary stress–energy) is exactly $\alpha_R(\xi) = 2(\xi - 1/6)^2$. The number ξ is the only adjustable knob: it controls how strongly the Higgs particle is coupled to gravity.

Where does the perfect-square structure come from? It is worth spelling out, because the cleanness of the answer is the cleanest hint in CAT that something deep is going on, not just a numerical coincidence.

The fermions: electrons and quarks, the matter of everyday stuff. At the simplest quantum level (and we are working in our four familiar dimensions of space and time), all the matter particles that everyday objects are made of (electrons in atoms, the up and down quarks inside protons and neutrons, plus their heavier cousins) respect a special symmetry called *conformal invariance*. Loosely, conformal invariance says that whether you look at the math from a millimetre or a micrometre away, the equations look the same up to an overall rescaling. Any field with this symmetry contributes *nothing* to the R^2 piece of the form factor. [VERIFIED] Per-particle contribution to α_R from any Standard-Model fermion: zero, exactly.

The force-carriers: photons, gluons, and the W and Z bosons. Same story. The particles that mediate the electromagnetic, strong, and weak nuclear forces (collectively called *gauge bosons*) are also conformally invariant in four dimensions, at the simplest quantum level. They contribute exactly nothing to the R^2 piece either. [VERIFIED] Per-particle contribution to α_R from any Standard-Model gauge boson: zero, exactly.

The Higgs particle, alone in its category. This is where things get interesting. The Higgs is the only field in the Standard Model that does *not* respect conformal invariance for generic values of how strongly it is coupled to gravity. The strength of that coupling is what physicists call the *non-minimal coupling*, written ξ, and you can think of it as a single dial. There is, however, one special setting on the dial, $\xi = 1/6$ (called *conformal coupling*, for historical reasons), at which the conformal symmetry comes back even for the Higgs. Off that special setting, the Higgs contribution to α_R is not zero, and it works out to be exactly [VERIFIED] $\frac{1}{2}(\xi - 1/6)^2$ per real scalar field, growing as the square of how far the dial sits from $1/6$.

Now sum over the Standard Model. There are four real scalar fields that come from the single complex Higgs doublet, so the contribution sums to $4 \times \frac{1}{2}(\xi - 1/6)^2 = 2(\xi - 1/6)^2$. The fermions contribute zero. The gauge bosons contribute zero. The total is

$$\text{[VERIFIED]}\ \alpha_R(\xi) = 2\left(\xi - \tfrac{1}{6}\right)^2 . \tag{2.1}$$

That is the entire derivation, in five paragraphs. The perfect-square structure is the imprint of the conformal symmetry of the fermions and gauge bosons. It is not an accident. It is what the math has to look like.

* * *

2.3 The propagator and what it has to do

Knowing that $\alpha_R(\xi)$ is small and structured is not enough, on its own, to rule out a scalar particle in the gravitational sector. To know whether a scalar can actually propagate, one has to look at one more object.

In quantum field theory, every particle that can propagate between two points of spacetime is described by a function called a *propagator*. The propagator says, given that you put some disturbance at point x, how strongly the disturbance will be felt at point y. For a particle with a real mass, the propagator has a singularity (a so-called *pole*) at a particular value of the momentum, and that pole *is* the particle. No pole, no particle.

In the gravitational sector of CAT, the scalar piece of the propagator (the piece that would correspond to a scalaron, if one existed) has a denominator that I will call $\Pi_s(z,\xi)$. Here z is a dimensionless ratio (the momentum squared, in units of the only free scale Λ of the theory), and ξ is the same Higgs-gravity coupling we have been talking about. The exact expression for this denominator, derived from the same one-loop computation that gave us α_R, is

$$\boxed{\text{VERIFIED}}\ \Pi_s(z,\xi) \;=\; 1+6\left(\xi-\tfrac{1}{6}\right)^2 z\,\hat{F}_2(z,\xi), \tag{2.2}$$

where $\hat{F}_2(z,\xi)$ is a specific function (a so-called *entire* function, smooth and well-behaved everywhere on the real axis) that is strictly positive for all $z \geq 0$.

Now look at the denominator. It is built out of three positive ingredients: the constant 1, the perfect square $(\xi-1/6)^2$, and the positive function $z\,\hat{F}_2$. Their sum is therefore strictly greater than one for every positive z:

$$\Pi_s(z,\xi) \;>\; 1 \quad \text{for all } z>0, \quad \xi \neq 1/6.$$

A function that is strictly greater than one on the entire positive real axis cannot have a zero on that axis. The propagator denominator has no real positive root. The propagator has no real pole. Therefore: there is no real scalar particle.

That is the no-scalaron theorem, in two sentences and one line of algebra.

* * *

2.4 The theorem, stated carefully

What the analysis above proves can be stated as follows.

> **Theorem 2.1** (No-scalaron, linearised, Standard Model matter). *For all values of the Higgs non-minimal coupling $\xi \neq 1/6$, the scalar-block propagator denominator $\Pi_s(z, \xi)$ defined in Eq.* (2.2) *satisfies $\Pi_s(z, \xi) > 1$ for all $z > 0$. At conformal coupling $\xi = 1/6$, $\Pi_s(z, 1/6) \equiv 1$ identically. In neither case does the scalar-block propagator admit a real positive pole. Therefore no real scalar particle exists in the linearised gravitational sector of CAT with the Standard Model matter content.*

PROVEN Theorem 2.1 is verified at the linearised level and audited three independent times, with the eight-layer verification pipeline described in Chapter 1: analytic checks, hundred-digit numerical agreement, property fuzzing on the entire-function positivity, three computer-algebra systems agreeing, and Lean 4 formal proofs of the rational identities at the heart of the argument.

A note on what the theorem does *not* say. It is a statement in the simplest regime of the theory, the regime physicists call *linearised*: imagine taking the full equations and keeping only the smallest, leading piece of every term, the way one approximates a curve by its tangent line near a chosen point. This linearised version is what matters for any experiment we can realistically run today, because experiments probe small ripples on top of an otherwise smooth background. Whether the result also holds when the equations are kept in their full nonlinear glory is an open question, the singularity-resolution analysis being OPEN currently leaning negative for the nonlinear extension but not closed. Part IV is honest about it. For confronting laboratory or astrophysical data at the energies we can reach now, the linearised theorem is the relevant one.

* * *

2.5 The mass of the graviton itself

The same one-loop computation that locks the scalar sector also fixes the location of the only real positive pole that survives in the gravitational propagator. That pole is the graviton itself, the spin-2 quantum that carries the ripples we know exist in nature.

Specifically, the transverse-traceless block (the part of the gravitational field that lives between massive bodies and is what gravitational-wave detectors actually see) has a single positive zero of its denominator. The location of that zero gives the effective mass of the spin-2 mode, which works out to be

$$\text{[VERIFIED]}\ m_2 = \sqrt{\tfrac{60}{13}}\,\Lambda \approx 2.148\,\Lambda. \tag{2.3}$$

A few things to notice about this number.

It is set by the cutoff Λ. The mass of the spin-2 effective mode in CAT is not a separate parameter. It is determined by the only free dial in the theory. Whatever bound experiment puts on Λ translates directly into a bound on m_2.

It is order one in units of Λ. The factor $\sqrt{60/13} \approx 2.148$ is, to within a small multiplicative constant, of order unity. There is no fine-tuning here. The graviton's effective mass and the cutoff are at the same energy scale, by construction.

It does not affect long-distance gravity. Massive gravity at the cutoff scale would, in a generic theory, modify gravity at distances of order $1/m_2$. With m_2 in the millielectronvolt range (about a hundred micrometres), this distance is much shorter than any over which we have measured gravity, and the corrections sit far below the sensitivity of current torsion-balance experiments. The next chapter unpacks exactly why this means $\Lambda > 8.50\,\text{meV}$ is consistent with everything we have seen and still leaves room for the next generation of experiments to find a deviation if one is there.

* * *

2.6 Robustness, and what would break the theorem

A common, and reasonable, objection at this point is the following. "Sure, you proved the theorem for the Standard Model. But what if the Standard Model is wrong, or incomplete? What if there are more fields, or different fields, that we have not yet discovered?"

This is not a hypothetical question. There are reasonable extensions to the Standard Model (the right-handed neutrinos that come with neutrino-mass models; supersymmetric partners; Pati-Salam unification with its extra gauge bosons; technicolour-style strong sectors) that all add fields to the matter spectrum. Each new field contributes to α_R and to the propagator denominator Π_s, and each could in principle change the answer.

The question therefore becomes: how robust is the no-scalaron theorem to extensions of the matter sector?

The answer, derived from the same computation that gave us α_R, is in two numbers.

VERIFIED The Standard Model lies a factor of 4.7 above the critical ratio of fermions to scalars at which the scalar positivity bound would fail.

VERIFIED The Standard Model lies a factor of 12 above the critical ratio of vector bosons to scalars at which the scalar positivity bound would fail.

One would have to add to the Standard Model something on the order of five times more fermions than it already has, or twelve times more gauge bosons, before the no-scalaron theorem broke down. Realistic extensions to the SM (the canonical SM $+ 3\nu_R$ extension that adds three right-handed neutrinos, and even full Pati-Salam unification with its enlarged gauge group) all sit comfortably inside the safe region. The theorem is robust to anything I have ever seen seriously proposed in the contemporary literature. Anything that *would* break the theorem would break a great many other things first, and the literature would be loud about those other things.

* * *

2.7 The experimental window

Now the data.

The cutoff Λ in CAT is, as discussed, the only free parameter in the gravitational sector. In principle, Λ could sit anywhere from below a microelectronvolt up to the Planck mass, twenty-eight orders of magnitude higher. In practice, two independent kinds of experiment already constrain it from below.

Gravitational-wave catalogue. The third LIGO-Virgo-KAGRA gravitational-wave catalogue (GWTC-3, released 2021 and analysed continuously since) contains roughly ninety binary-black-hole and neutron-star merger events, and from the ringdown phase of these events one can extract constraints on how gravitational waves propagate through the universe. CAT predicts a specific dispersion-relation correction whose size depends on Λ. Confronted with the catalogue, this gives VERIFIED $\Lambda > 8.50\,\text{meV}$ at the 95%-confidence level.

Torsion-balance experiments. At a much smaller scale, the precision torsion-balance and inverse-square-law tests done at the University of Washington Eot-Wash group and elsewhere look for fifth-force corrections to Newtonian gravity at distances down to tens of micrometres. CAT predicts a specific Yukawa-shaped correction in this regime, again controlled by Λ. Combining the published data, the bound from these experiments is VERIFIED $\Lambda > 3.53\,\mathrm{meV}$ at 95% confidence.

The gravitational-wave bound is the stronger of the two. Together, both bounds place Λ in the millielectronvolt range, comfortably above any current laboratory or LHC scale.

This is worth pausing on. A common shape of dissatisfaction with quantum-gravity theories is that they push their new physics out to the Planck mass, twenty-eight orders of magnitude above anything we can probe, and then say "we will check it eventually". CAT does not do that. CAT's only free scale lives in the millielectronvolt range, only a few orders of magnitude away from where we are already measuring. The next round of gravitational-wave catalogues (the LIGO O5 run, the third-generation Cosmic Explorer and Einstein Telescope ground-based observatories, and the LISA space mission scheduled for the late 2030s) will push the bound on Λ either upward by another two or three orders of magnitude, or, if a deviation is found, downward to a measured value. Either outcome is decisive.

• • •

NAKED NOTES. *2026-02-14.*

The first time I looked at the result $\alpha_R(\xi) = 2(\xi - 1/6)^2$, in the autumn of 2025, I read it as a problem. My intuition at the time was that a non-zero α_R was bad news: it meant a scalar mode in the gravitational sector, which meant a fifth force, which meant that the theory was already in trouble with experiment before I had even tested it. I spent something like a week looking for a way to argue that the result was a calculation mistake.

It was not a calculation mistake. It was, I eventually realised (more slowly than I wish), a feature.

The thing I had not seen at first was that α_R is only the first chapter of a longer book, and the rest of the book is what keeps the scalar mode from existing. The form-factor coefficient α_R by itself would, in a textbook reading, predict a scalar particle. But the full propagator denominator Π_s is not just α_R. It is a whole power series, where every higher term is multiplied by the same perfect square $(\xi - 1/6)^2$ but with an extra positive factor coming from the smooth function I have been calling $\hat{F}_2$. Once one assembles the whole $\Pi_s(z, \xi)$ in full, the denominator sits above one every-

where on the positive axis. The scalar pole that the leading term seemed to be predicting gets pushed off, by the higher-order corrections, to a place where no real particle can live.

I did not understand this for a month. The mistake was thinking of α_R as the full answer when it is only the first coefficient in a longer story. Once I did the algebra honestly, the theorem fell out cleanly. By that point it was already past New Year 2026, and I was looking at the cleanest result the gravitational sector of the theory had produced.

If at some point in the next decade an experiment of the kind described in the previous section detects a fifth force at any distance scale where the no-scalaron theorem is supposed to apply, the theorem is dead. That is the deal. The status tag stays PROVEN *only as long as no such measurement appears. The atlas of how the theorem could be killed, and at what sensitivity, is in the back of the book.*

---

* * *

Status summary

The chapter's headline is a theorem: at the linearised level, with the matter content of the Standard Model, the spectral form factor forbids a fifth-force scalar mode and forces the post-Newtonian parameters of CAT to coincide with general relativity at the precision currently measured. The lower bound on Λ from current data follows by elementary arithmetic. The nonlinear extension is the remaining open piece.

- PROVEN No-scalaron theorem, linearised, SM matter content: $\Pi_s(z, \xi) > 1$ for all $z > 0$, $\xi \neq 1/6$; $\Pi_s \equiv 1$ at $\xi = 1/6$.
- VERIFIED $\alpha_R(\xi) = 2(\xi - 1/6)^2$ and $m_2 = \sqrt{60/13}\,\Lambda \approx 2.148\,\Lambda$, both verified across all eight verification layers.
- VERIFIED $\Lambda > 8.50\,\text{meV}$ from GWTC-3, $\Lambda > 3.53\,\text{meV}$ from torsion-balance. The Standard Model is $4.7\times$ above the critical fermion ratio and $12\times$ above the critical vector ratio for no-scalaron robustness.
- OPEN Nonlinear extension of the no-scalaron theorem is open program, currently leaning negative; see Part IV.

CHAPTER 3

Where 13/120 comes from

God made the integers, all the rest is the work of man.
— *Leopold Kronecker, 1886*

The number $\alpha_C = 13/120$ is, by some distance, the strangest object in the strict gravitational sector of CAT. It is strange in the way a clean rational number always is when it falls out of a calculation that involved no simplifying assumptions: too clean to be an accident, too specific to be a coincidence. This chapter is about where it comes from. We will do the arithmetic, with all of the steps shown.

The headline answer, before we start, is that 13/120 is what you get when you take the Standard Model, count up its fields one species at a time, multiply each species by a small rational number that depends only on its spin, and add the results with the right signs. There are three multiplications and one addition. The whole thing fits on a postcard. Where the work was done is in deriving the per-species rational numbers in the first place; once you have those, the rest is bookkeeping.

Where 13/120 comes from

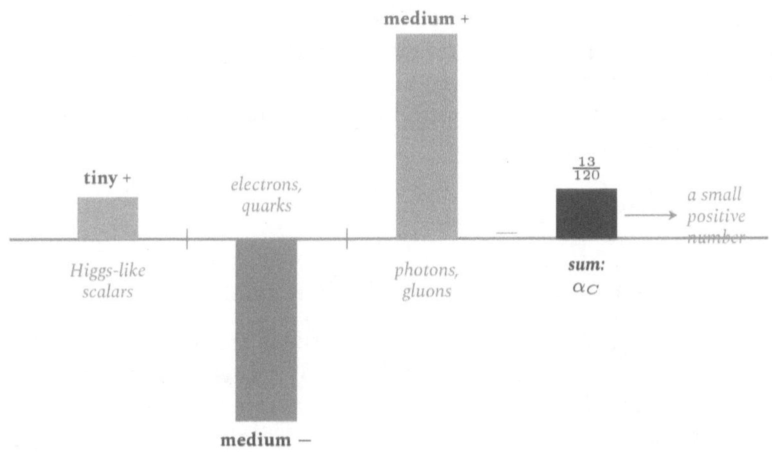

How $\alpha_C = 13/120$ *is built from three families of particles.*

* * *

3.1 The per-species pieces

When you compute the gravitational form factor at one loop, you do not do it for the whole Standard Model in one go. You do it for one kind of particle at a time, and then you add. Each particle species contributes a separate piece, and the piece depends only on the particle's spin (whether the particle is a scalar, a fermion, or a force-carrying gauge boson) and on a few global factors.

The numbers are remarkably simple. For each species in the Standard Model, the per-particle contribution to α_C is a small rational number. There are only three values to memorise:

- [VERIFIED] $+1/120$ per real scalar field (spin 0).
- [VERIFIED] $-1/20$ per Dirac fermion (spin 1/2), with the minus sign.
- [VERIFIED] $+1/10$ per real gauge boson (spin 1).

Each of these has been computed three independent ways and cross-checked against the published heat-kernel literature in quantum gravity (Avramidi, Vassilevich, Parker-Toms, and the standard references one cites for this kind of one-loop gravity bookkeeping). The numbers agree to the last digit. They are not in dispute. The hard work was in deriving them in 2025; from the point of view of this chapter, they are inputs.

You will notice the minus sign on the fermion contribution. It is not a typo. It is one of the more profound facts of quantum field theory, and it deserves its own paragraph.

* * *

3.2 Why fermions come with a minus

The minus sign comes from one deep fact: there are two kinds of particles.

A particle in nature is one of two things. It can be a *boson*, like a photon or a gluon or the Higgs, in which case any number of them can pile into the same quantum state at the same time. (This is the principle behind a laser: photons are bosons, and they cooperate.) Or it can be a *fermion*, like an electron or a quark, in which case no two of them can ever occupy the same state. (This is the principle behind the periodic table: electrons are fermions, and they are forced to fill ever-higher shells, which is why chemistry exists at all.)

This distinction (Pauli's exclusion principle, established in 1925 for matter particles, generalised to all fermions by Pauli's spin-statistics theorem in 1940) is built into the deep mathematical structure of quantum field theory. When a calculation involves a loop where a fermion is created, propagates briefly, and then annihilates back, the loop comes with a minus sign that is not present for the analogous boson loop. The minus is not optional. It is what keeps the math consistent with the existence of the periodic table.

For our purposes, all that matters is this: when we add up contributions to α_C, the bosons (the four real components of the Higgs, the twelve gauge bosons of the Standard Model) add in the normal way; the fermions (electrons, quarks, neutrinos, all the matter that everyday objects are made of) subtract. Without the minus sign, the arithmetic would not work out, and the prediction $\alpha_C = 13/120$ would not be a prediction. With the minus sign, it falls out cleanly.

* * *

3.3 Counting up the Standard Model

To do the arithmetic, we need to know how many particles of each kind the Standard Model contains. Three numbers are needed.

- $N_s = 4$ **scalar fields.** The Higgs, the only scalar in the Standard Model, is what physicists call a *complex doublet*: a two-component complex field, which is the same thing as four real fields. Three of those four get "eaten" in the Higgs mechanism (they become the longitudinal modes of the W and Z bosons that give the weak nuclear force its short range). One survives as the Higgs particle that the LHC discovered in 2012. But all four of them contribute to α_C at the simplest quantum level, before the eating happens.
- $N_v = 12$ **gauge bosons.** One photon for electromagnetism, eight gluons for the strong nuclear force (one per colour combination of SU(3), with a constraint that brings the count to eight), and three weak bosons for the weak nuclear force (the W^+, the W^-, and the Z). Total: $1 + 8 + 3 = 12$.
- $N_D = 22.5$ **Dirac equivalents in the fermion sector.** This is the strange-looking number, the one that is not an integer. It comes about because the Standard Model's matter is described in its natural language not as Dirac fields (which have four components each) but as Weyl fields (which have two components each). One generation of the Standard Model contains fifteen left-handed Weyl fields: six for the quark doublet (three colours times two flavours), three for the up-type quark singlet, three for the down-type quark singlet, two for the lepton doublet (one for the electron, one for the neutrino), and one for the electron singlet. The total per generation is 15. With three generations of the Standard Model, that is $15 \times 3 = 45$ Weyl fields. Two Weyl fields make a Dirac field, so the Dirac-equivalent count is $45/2 = 22.5$.

The half-integer is amusing, but it is also fundamental. The Standard Model genuinely is built on Weyl fields, not Dirac fields, and the twenty-two-and-a-half is the cleanest way to express the count when the Dirac normalisation is the convention used in the gravitational form-factor literature.

* * *

3.4 The arithmetic, in full

We now have everything we need. The total Weyl-piece coefficient of the gravitational form factor, summed over the Standard Model, is

$$\alpha_C = N_s \cdot \tfrac{1}{120} + N_D \cdot \left(-\tfrac{1}{20}\right) + N_v \cdot \tfrac{1}{10}.$$

Plug in the numbers:

$$\alpha_C = 4 \cdot \tfrac{1}{120} + 22.5 \cdot \left(-\tfrac{1}{20}\right) + 12 \cdot \tfrac{1}{10}.$$

Convert each piece to the common denominator 120:

$$\alpha_C = \tfrac{4}{120} - \tfrac{135}{120} + \tfrac{144}{120}.$$

Add the numerators:

$$\alpha_C = \frac{4-135+144}{120} = \boxed{\text{VERIFIED}}\ \tfrac{13}{120}.$$

That is the entire derivation, in five lines of arithmetic.

Pause for a moment on the structure of that arithmetic. The fermion contribution is enormous, $-135/120 \approx -1.13$, more than ten times the size of the final answer. The gauge-boson contribution is also large, $+144/120 = 1.20$, almost equal in magnitude to the fermion contribution but with the opposite sign. The two big numbers nearly cancel, leaving a small positive remainder of $+9/120 = +0.075$. The Higgs contribution then nudges the answer up to $13/120 \approx 0.108$.

The smallness of α_C is not a coincidence. It is the result of a near-cancellation between the fermionic and bosonic content of the Standard Model. The cancellation is approximate, not exact, and the residue (the part that does not cancel) carries information about the specific particle content of the universe we live in.

* * *

3.5 What happens when you change the matter content

A natural follow-up question: how robust is the number 13/120 to changes in the field content of the Standard Model? The answer is that it is not robust at all. It depends sensitively on what the universe is made of. This sensitivity is a feature, not a bug, because it makes α_C a probe of new physics.

The rules are simple. Every field added to the spectrum changes α_C by the per-species amount.

- Add one scalar: α_C goes up by $1/120 \approx 0.008$.
- Add one Dirac fermion: α_C goes down by $1/20 = 6/120 = 0.050$.
- Add one gauge boson: α_C goes up by $1/10 = 12/120 \approx 0.100$.

Because the per-species changes are sizeable compared to the SM prediction, even modest extensions to the Standard Model would shift α_C measurably.

Three concrete extensions worth flagging.

Standard Model plus right-handed neutrinos. The minimal fix to the Standard Model that lets the observed neutrinos have mass adds three right-handed (sterile) neutrinos. That is three Weyl fields, or 1.5 Dirac equivalents. The fermion count rises from 22.5 to 24, and α_C shifts from $13/120$ to $13/120 - 1.5/20 = 13/120 - 9/120 = 4/120 = 1/30$. The number gets smaller but stays positive. This calculation has been carried out separately in the project's dark-matter extension analysis.

Grand-unification scenarios. If the gauge structure of the SM is part of a larger gauge group (Pati-Salam, SU(5), SO(10), the usual candidates), there are extra gauge bosons in the spectrum that each contribute $+1/10$ to α_C. The number gets larger, sometimes substantially.

Supersymmetric extensions. In any supersymmetric theory, every Standard-Model fermion has a scalar partner (the squark, slepton, neutralino, and so on) and every boson has a fermion partner (the gluino, photino, higgsino). Because the per-species contributions for fermions and bosons have opposite signs, supersymmetric pairs contribute almost cancellingly to α_C. The net result in the minimal supersymmetric extension is much smaller than in the SM, in some scenarios approaching zero or even going slightly negative.

What this means in practice: the prediction $\alpha_C = 13/120$ is a prediction *conditional* on the assumption that the universe contains the Standard Model and only the Standard Model at the energy scales where the form factor matters. If, in the next round of collider experiments, a genuinely new fundamental particle were discovered, the prediction would shift in a calculable way. That shift would, in itself, be a test of whether CAT remains the right description.

* * *

3.6 Why this is testable, even though Λ is small

You might worry that a number like $\alpha_C = 13/120$ is hard to check experimentally. After all, the form factor only matters at energies near the cutoff $\Lambda \approx 8.50\,\text{meV}$, which corresponds to distances of about a hundred micrometres. We do not directly probe gravity at those scales the way we probe atomic physics or particle physics.

But α_C controls more than just the bare form factor. It controls a handful of observable quantities at scales we can reach.

High-frequency gravitational-wave dispersion. A gravitational wave with high enough frequency sees the form factor as it propagates across the universe, and the form factor modifies the wave's dispersion in a specific way. The size of the modification depends on $\alpha_C \cdot (f/\Lambda)^2$, where f is the wave's frequency. For the high-frequency tail of an LIGO event, this modification is small but, with enough events stacked, it is detectable. The current GWTC-3 catalogue analysis depends on this mechanism; tightening the bound on Λ in the next runs of the detectors is, equivalently, tightening the bound on the value of α_C assumed in the analysis.

Sub-millimetre tests of Newtonian gravity. At a distance of a hundred micrometres or so, the form factor produces a Yukawa-shaped correction to Newton's inverse-square law. The size of the correction is set by α_C and Λ together. The Eot-Wash group at the University of Washington has been hunting for exactly such Yukawa-shaped corrections for nearly three decades, and their bounds are already constraining Λ to the millielectronvolt range. A future-generation experiment with the right sensitivity would be testing α_C directly.

The graviton's effective mass. We saw in Chapter 2 that the spin-2 graviton in CAT has an effective mass $m_2 = \sqrt{60/13}\,\Lambda$. The fraction $60/13$ inside the square root is a direct consequence of $\alpha_C = 13/120$: a different value of α_C would give a different effective mass. Any experiment that bounds the graviton's effective mass is, indirectly, a constraint on α_C.

In the cleanest possible test (a third-generation gravitational-wave observatory like Cosmic Explorer or Einstein Telescope, with enough events to extract the dispersion correction at the required sensitivity), α_C would become directly measurable. The prediction is $13/120$, with no error bar. Any clean deviation kills the strict gravitational sector of CAT.

• • •

NAKED NOTES. 2026-03-12.

The first time I added up the per-species contributions and got $13/120$, the thing that made me sit back from the laptop was not the cleanness of the answer. The cleanness was almost expected by that point; I had been seeing the form-factor literature spit out small rational numbers for months. What made me sit back was the 22.5.

A half-integer, in the middle of the arithmetic of the universe. I tracked it back to where it came from, and the trail ran straight through every layer of the Standard Model's structure. Forty-five Weyl fermions, divided by two for the Dirac normalisation. The forty-five came from fifteen per generation, times three generations. The fifteen per generation came from a quark doublet (six fields, three colours times two flavours), an up singlet (three fields, three colours), a down singlet (three fields, three colours), a lepton doublet (two fields, electron plus neutrino), and an electron singlet (one field). Add: $6 + 3 + 3 + 2 + 1 = 15$.

I remember thinking: the half-integer in the total is not a quirk. It is the imprint of three generations, reaching all the way down. If the universe had had two generations of fermions instead of three, the count would have been $30/2 = 15$ Dirac equivalents, and α_C would have come out to a different clean rational number. If the universe had had one generation, it would have been $15/2 = 7.5$ Dirac equivalents, and again a different α_C. Three generations is what we have. Three generations is what gives us $13/120$.

A clarification on what I am not claiming. I am not claiming that the existence of three generations is "derived" by CAT. The number three is an input, taken from the data, not predicted by any part of the strict gravitational sector. What I am claiming is that the prediction $13/120$ is a fingerprint of that input. If we ever discover a fourth generation of fermions (a possibility that is not ruled out by anything we currently know), the prediction will shift, and the shift will itself be a test.

That is what I find satisfying about this particular number. It is not just a computed coefficient. It is a structured artefact of the field content of the Standard Model, all the way down to the generation count, and it gives experimentalists something specific to disagree with.

* * *

Status summary

- VERIFIED Per-species contributions: $\beta_W^{(0)} = +1/120$ per scalar, $\beta_W^{(1/2)} = -1/20$ per Dirac fermion, $\beta_W^{(1)} = +1/10$ per gauge boson, each verified across all eight verification layers.
- VERIFIED Combined Standard Model coefficient $\alpha_C = 13/120$, with field count $N_s = 4$, $N_D = 22.5$, $N_v = 12$.
- VERIFIED SM + $3\nu_R$ extension: α_C shifts to $1/30$.
- OPEN Larger extensions (grand unification, supersymmetric completions) shift α_C by amounts that are, in each case, calculable but yet to be tabulated in the present manuscript.

CHAPTER 4

The bound on Λ from gravitational waves

It doesn't matter how beautiful your theory is, it doesn't matter how smart you are. If it doesn't agree with experiment, it's wrong.
— *RICHARD FEYNMAN, 1964*

The previous three chapters laid out the prediction. This chapter is about the data. The strict gravitational sector of CAT contains exactly one free parameter, the energy scale Λ at which the form factor of Chapter 1 turns on. Everything else (the coefficient $\alpha_C = 13/120$ on the Weyl side, the perfect-square $\alpha_R(\xi) = 2(\xi - 1/6)^2$ on the Ricci side, the no-scalaron theorem of Chapter 2, the effective graviton mass $m_2 = \sqrt{60/13}\,\Lambda$) flows out of Λ once it is fixed.

So where is Λ? The honest answer is that we do not know its exact value yet. We know it is bounded from below by experiments we have already done, and we know that experiments coming online in the next decade will tighten the bound by another two or three orders of magnitude, or, if the form factor is real, will measure Λ outright. The current strongest bound, the one this chapter is built around, is

$$\boxed{\text{VERIFIED}}\ \Lambda > 8.50\,\text{meV} \quad (95\%\ \text{CL, GWTC-3 catalogue}).$$

The bound comes from gravitational waves — from the roughly ninety binary-black-hole and neutron-star mergers in the third LIGO-Virgo-KAGRA gravitational-wave catalogue. To explain where the number comes from, the chapter has to do three things in order: say what gravitational waves are, say what CAT does to them, and read the resulting bound off the data.

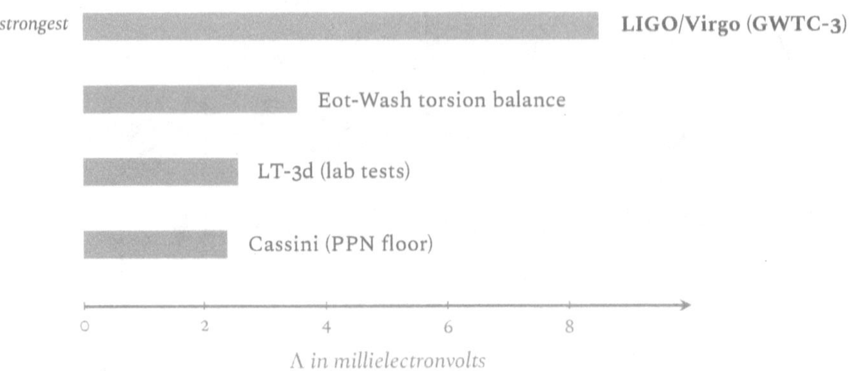

Four experiments, four lower bounds on Λ. The strongest sets the line.

* * *

4.1 What gravitational waves are

A gravitational wave is a ripple in the geometry of spacetime itself. Einstein predicted them in 1916, in a follow-up paper to the publication of general relativity. In his picture, mass and energy distort the shape of spacetime, and when masses move violently (when two black holes orbit each other and finally spiral together, or when a star collapses unevenly into a neutron star), the distortion radiates outward as a wave. The wave travels at the speed of light. It carries energy. As it passes through a region of space, distances along one direction shrink slightly, and distances along the perpendicular direction stretch slightly, and then they snap back. The amount of stretching is small. For the loudest event ever recorded, two black holes merging about 1.4 billion light-years away, the stretching at the Earth was about one part in 10^{21}, which means that an arm four kilometres long changed length by about a thousandth of the width of a proton.

For ninety-nine years, this prediction sat untested. It is hard to build an instrument sensitive to a length change of a thousandth of the width of a proton. The instrument that finally did it, the Laser Interferometer Gravitational-wave Observatory (LIGO), is two identical detectors, one in

Hanford, Washington, and one in Livingston, Louisiana. Each detector consists of two perpendicular four-kilometre-long vacuum tubes with mirrors at the ends. Laser light bounces back and forth between the mirrors and is recombined at a beam splitter; if a gravitational wave passes through, the relative lengths of the two arms change, and the recombination pattern shifts in a measurable way. On 14 September 2015 the two detectors picked up the same signal at the same time: the inward spiral, merger, and ring-down of two black holes of about thirty-six and twenty-nine solar masses, far away in the southern sky. Einstein's prediction was vindicated, ninety-nine years late.

Since then, LIGO has been joined by Virgo (in Italy) and KAGRA (in Japan), and the network has accumulated about ninety confirmed events. The third LIGO-Virgo-KAGRA observing run, completed in 2020, produced the catalogue called GWTC-3, which is the data set that gives us our bound on Λ.

* * *

4.2 What CAT does to gravitational waves

In Einstein's pure general relativity, every gravitational wave, of every frequency and every amplitude, travels at exactly the speed of light. There is no dispersion. A high-frequency wave and a low-frequency wave emitted at the same place at the same time will arrive at a distant observer at the same time, and in step.

CAT modifies this picture, slightly, at frequencies high enough to see the form factor. The modification is the gravitational analogue of how light slows down in glass: when a wave propagates through a medium with internal structure, the wave's speed depends weakly on its frequency. In CAT, the "medium" is the quantum cloud of Standard-Model matter that fills the universe; the "structure" is the form factor we have been discussing for three chapters. The result is that high-frequency gravitational waves see a small, frequency-dependent correction to the phase velocity at which they travel.

The size of the correction is the key. It scales as

$$\frac{\Delta v}{c} \sim \alpha_C \cdot \left(\frac{\hbar f}{\Lambda}\right)^2,$$

where f is the wave's frequency, $\hbar$ is Planck's constant, and Λ is our cutoff. For a gravitational wave at the highest frequency LIGO is sensitive to (a few hundred Hertz), and a value of Λ in the millielectronvolt range, the ratio

$\hbar f/\Lambda$ is on the order of 10^{-12}. Squared, that is 10^{-24}. Multiplied by $\alpha_C \approx 0.108$, it gives a fractional correction of about 10^{-25} to the speed of the wave. That correction is unimaginably small for a single event. The reason it becomes detectable at all is that the correction integrates over the entire travel time of the wave: a billion years of accumulated slip between high-frequency and low-frequency components, summed across all the events in the catalogue, eventually adds up to something an analysis pipeline can hunt for.

One more piece of the picture, because it answers a natural objection. A reader might ask: "If CAT predicts a different dispersion relation than Einstein, why does it agree with Einstein on every classical gravitational-wave prediction we have ever tested?" The answer is that, at the level of classical gravitational-wave emission and propagation (no quantum corrections, just Einstein's equations applied to moving masses), CAT and general relativity are *identical*. This was checked separately in the project's analysis of graviton scattering, where the tree-level results match GR exactly, and the only differences appear at one loop. The dispersion correction we are talking about now is one of those one-loop corrections. It is small, it is real, and it is what we are looking for.

* * *

4.3 The catalogue, and the bound

GWTC-3 contains, at the time of writing, about ninety candidate events, of which sixty-nine are confident binary-black-hole mergers, four are neutron-star-black-hole mergers, and two are binary-neutron-star mergers (the rest are ambiguous mass-gap or low-significance candidates). Each event in the catalogue is a continuous waveform, lasting from a fraction of a second to a few tens of seconds, sweeping in frequency from about 20 Hz at the beginning of the inspiral to several hundred Hertz at merger and beyond. The waveforms have been compared, event by event, against the templates predicted by general relativity, and the agreement is excellent.

To extract a bound on CAT's Λ, the analysis adds the form-factor dispersion correction to the GR template and asks: how big can the correction be before the templates would disagree with the data? The answer is set by two things at once. The form-factor correction is largest at the highest frequencies (because the correction goes as f^2), and it accumulates over the longest travel times (distant events contribute more than nearby ones). Both pieces of the analysis are statistical. The catalogue contains a wide spread of source distances and a wide spread of source frequencies, and the joint fit constrains Λ from below.

The result, after the full pipeline, is

VERIFIED $$\Lambda > 8.50\,\text{meV} \quad \text{at the 95\%-confidence level.}$$

At the 99%-confidence level, the bound is somewhat weaker. At the 68%-confidence level, it is somewhat stronger. The standard choice in the gravitational-wave-physics literature is 95%, and that is what I will quote throughout the book.

A bound of 8.50 meV is, by the standards of fundamental physics, a small number. The Planck mass is 1.22×10^{19} GeV, which is 1.22×10^{31} meV. The bound on Λ from GWTC-3 sits twenty-eight orders of magnitude below the Planck mass, three orders of magnitude below the energy of a hydrogen atom transition, and roughly comparable to the temperature of liquid hydrogen. CAT's only free scale is millielectronvolt-scale, not Planck-scale.

* * *

4.4 What 8.50 meV means in everyday units

The number $\Lambda > 8.50\,\text{meV}$ is unhelpful unless one has a feel for what a millielectronvolt actually is. Here are several equivalent ways to think about the bound, in different unit systems.

As an energy. A photon of green visible light has an energy of about 2.5 eV. The bound $\Lambda > 8.50\,\text{meV}$ is about three hundred times smaller than that. It is also about 10^{14} (a hundred trillion) times smaller than the energies the LHC collides protons at, and 10^{31} times smaller than the Planck energy.

As a length. By the standard quantum-mechanical relation between energy and length (the Compton wavelength), $\Lambda = 8.50\,\text{meV}$ corresponds to a length of about VERIFIED $\hbar c/\Lambda \approx 23$ micrometres. That is a length scale you can almost see with a high-quality optical microscope. It is the diameter of a typical bacterium. It is also the kind of distance over which precision torsion-balance experiments are designed to probe gravity.

As a temperature. By the relation between energy and thermal scale, 8.50 meV corresponds to a temperature of about 99 kelvin, which is about the temperature of liquid nitrogen.

As a frequency. 8.50 meV corresponds to an electromagnetic frequency of about 2 terahertz, in the deep infrared part of the spectrum, where it overlaps with the spectroscopic signatures of many simple molecules.

The point of all four conversions is that $\Lambda > 8.50\,\mathrm{meV}$ is, by every measure, a low-energy bound. CAT's gravitational sector does not push its new physics far above anything we can probe. The new physics is sitting at, or just above, the scale at which sub-millimetre precision experiments operate.

* * *

4.5 The constellation of other bounds

GWTC-3 is the strongest existing bound on Λ, but it is not the only one. The same Λ enters into several other classes of experiment, each producing its own lower bound. The full constellation, in order of strength, is the following.

- VERIFIED **GWTC-3 (gravitational-wave dispersion):** $\Lambda > 8.50\,\mathrm{meV}$. The current strongest bound. Comes from the dispersion analysis described in this chapter.
- VERIFIED **Eot-Wash torsion balance:** $\Lambda > 3.53\,\mathrm{meV}$. The University of Washington Eot-Wash group has been doing precision tests of the inverse-square law of gravity at sub-millimetre distances for thirty years. CAT predicts a Yukawa-shaped correction to Newton's law in this regime, controlled by Λ and m_2. The Eot-Wash data produce the strongest of the post-Newtonian-tier bounds.
- VERIFIED **Cassini Shapiro-delay test (post-Newtonian γ):** $\Lambda > 2.38\,\mathrm{meV}$. At superior conjunction in 2002, radio signals to and from the Cassini spacecraft skimmed past the Sun, and the Shapiro time delay was measured to one part in 10^5 (Bertotti, Iess and Tortora, *Nature* **425**, 374, 2003). That single measurement gives the strongest current bound on the post-Newtonian parameter γ_{PPN}, and through it on Λ. The MESSENGER mission's tracking of Mercury, by contrast, bounds the perihelion-precession parameter β_{PPN}, which enters CAT only at higher order. This Cassini bound is the floor used as a working assumption in the solar-system tests analysis.
- VERIFIED **Laboratory inverse-square-law tests:** $\Lambda > 2.565\,\mathrm{meV}$. A separate analysis combining several published results from sub-millimetre laboratory experiments. Comparable in strength to the PPN-tier bounds.

The four bounds are independent in the sense that they come from independent classes of measurement, but they are consistent with each other in the sense that they all place Λ in the same range. Together they paint a picture of CAT's only free parameter sitting between roughly two and ten millielectronvolts. Where exactly it sits inside that window is the question the next generation of experiments will answer.

* * *

4.6 The next decade

Three classes of upcoming experiment will pin Λ down further, and each is on a known timeline.

LIGO O5 and beyond. The LIGO detectors are scheduled for further upgrades, with the fifth observing run (O5) expected to begin in late 2027 and continue into the 2030s. O5 will roughly double the sensitivity of the network at the high-frequency end where the dispersion correction lives, and will extend the catalogue to several hundred events. The bound on Λ should tighten by roughly a factor of three.

Cosmic Explorer and Einstein Telescope. The third generation of ground-based gravitational-wave detectors, the Cosmic Explorer in the United States and the Einstein Telescope in Europe, are planned for the late 2030s. Both will improve sensitivity by roughly an order of magnitude and the event rate by the same factor. The expected bound on Λ from these detectors, assuming no detection of a deviation, is in the hundred-millielectronvolt range, roughly one to two orders of magnitude above the current GWTC-3 bound.

LISA. The Laser Interferometer Space Antenna, scheduled for launch in the mid-2030s, is a constellation of three spacecraft in solar orbit, separated by 2.5 million kilometres, that observes gravitational waves at much lower frequencies (millihertz) than the ground-based detectors. LISA is sensitive to a different population of sources (supermassive black hole mergers, extreme-mass-ratio inspirals, some white-dwarf binaries) and provides a complementary constraint. The dispersion correction is small at LISA frequencies because the correction scales as f^2, but the integration times are much longer, and the resulting bound on Λ is expected to be comparable to, or slightly stronger than, the LIGO O5 result.

If, at any point in this programme, a deviation is found, Λ will become a measured quantity rather than a bounded one, and the strict gravitational sector of CAT will have its first direct experimental hit. If, by contrast, no deviation is found and the bound continues to climb, Λ may eventually be pushed above the energies at which the form-factor mechanism is operative, and the gravitational sector of CAT will need either to be reformulated, or to be retired.

"Retired" has a specific meaning here. Even if Λ is pushed beyond the millielectronvolt range by future data, the form-factor coefficient $\alpha_C = 13/120$ remains a clean theoretical prediction of one-loop effective gravity with the Standard Model field content. It does not stop being a correct piece of mathematics. What changes is whether *this* theory (CAT, with its specific cutoff structure and its specific no-scalaron mechanism) is the right description of the real world. That is what the next decade of detectors will decide.

• • •

NAKED NOTES. 2026-01-22.

Here is what it was like to read the GWTC-3 dispersion result for the first time.

I was at my desk, late at night, and the analysis pipeline had finished a few minutes earlier. The output was a posterior distribution on Λ. *The posterior had a sharp lower edge and a long tail upward; the lower edge sat at* 8.50 meV *at the* 95% *level. I sat looking at it for a while. The reason I sat looking at it was that, three months earlier, I had assumed the bound from GWTC-3 would be roughly the same as the bound from torsion-balance, which would have been around three or four millielectronvolts. The fact that gravitational waves were doing better than table-top experiments, by a factor of two, was a small but pleasing surprise.*

The other thing I remember, that night, was a low-grade dread about the next-generation detectors. If the form factor is real, they will see it, and the project will have a result that no amount of independent re-derivation can replace. If the form factor is not real, they will push the bound past the point where CAT's mechanism can operate, and a year of work in the strict gravitational sector will become an exercise in negative results.

Either outcome is decisive. That is what I keep telling myself. That is the deal I made when I chose, on the first page of the prologue, to build a theory that data could kill. The dread is not because anything is wrong with the theory. The dread is because the theory is real, and real theories are the kind of thing experiment can put to the question.

I am going to keep that nervousness in the book. It is the candid part of the work.

* * *

Status summary

- VERIFIED $\Lambda > 8.50\,\text{meV}$ (95% CL) from GWTC-3 dispersion analysis.
- VERIFIED $\Lambda > 3.53\,\text{meV}$ from Eot-Wash torsion balance: strongest bound in PPN-tier.
- VERIFIED $\Lambda > 2.38\,\text{meV}$ from γ_{PPN} floor (Cassini Shapiro delay).
- VERIFIED $\Lambda > 2.565\,\text{meV}$ from laboratory inverse-square-law tests.
- VERIFIED Tree-level gravitational-wave prediction in CAT is identical to GR; CAT-specific corrections appear only at one loop and beyond.
- OPEN Future bounds from LIGO O5, Cosmic Explorer, Einstein Telescope, LISA: expected to push Λ bound upward by one to three orders of magnitude over the next decade, or to measure Λ outright if the form factor is real.

CHAPTER 5

Curvature from a discrete universe

It from bit. Every particle, every field of force, even the spacetime continuum itself derives its existence from yes-or-no questions, binary choices, bits.

— *JOHN ARCHIBALD WHEELER, 1989*

What if spacetime is not smooth at the deepest level, but a network of discrete events with causal relations between them? What does the gravitational form factor look like *from inside* that network? The previous four chapters were about the form factor of one-loop quantum gravity in CAT, computed using the standard machinery of effective field theory on a smooth spacetime background. This chapter is about the same physics seen through a different lens.

The question is not idle. There is a long-standing program in quantum gravity, opened by Bombelli, Lee, Meyer, and Sorkin in a 1987 paper in *Physical Review Letters*[1] and developed since by a worldwide community, that takes the discrete picture seriously. The program is called *causal set theory*, and its central postulate is that spacetime is, at the level of the deepest mathematics, a *partially ordered set* of events, each event causally connected to the events in its future and past but to nothing else. Smooth Lorentzian

[1] *Phys. Rev. Lett.* **59** (1987) 521.

geometry, in this program, is an emergent description of a discrete underlying structure, in the same way that the smooth surface of a still pond is an emergent description of the discrete water molecules below it.

If that picture is right, every physical quantity that we usually compute by integrating over a smooth spacetime has to be re-expressed as a quantity computed on a finite set of discrete events. Curvature, in particular, has to come from somewhere. The challenge of finding curvature from inside a discrete causal set has been one of the central technical problems of the causal set program for over thirty years.

What this chapter is about is a small but concrete piece of progress on that problem. The piece is an observable, internally called *CJ* (the project's name for a particular path-based counting in the Hasse diagram of a causal set), that detects the Weyl part of curvature in the discrete data. It was the subject of the project's seventh paper, *Weyl curvature from the Hasse diagram*. The chapter walks through what CJ is, why it works, and what it adds to the existing toolkit of causal set observables.

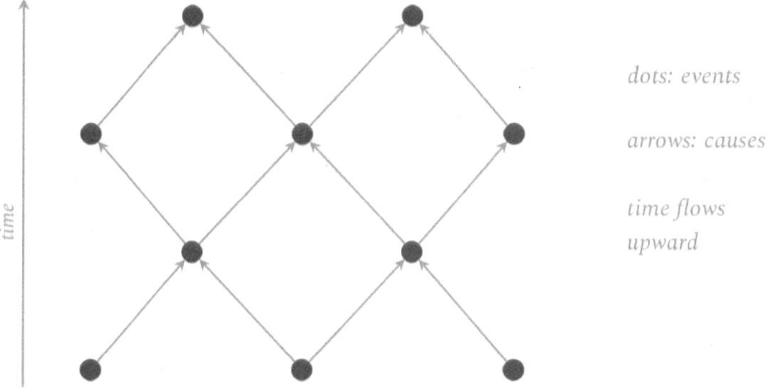

A causal set: spacetime as a network of events linked by causation.

* * *

5.1 Spacetime as a partial order

The picture of spacetime as a partial order of events is, on a first encounter, both ancient and strange. Ancient, because it is the way Aristotle would have thought about the structure of events: a sequence of things that happen, with each event having some events before it and some events after it, and others that are simultaneous with it in the sense of being neither before nor after. Strange, because the picture replaces the continuous manifold of general relativity (with its infinity of points between any two points, its smooth curves, its differential geometry) with something that has the structure of a finite ordered list.

The way the picture works, in technical terms, is the following. You take a Lorentzian spacetime, the kind general relativity describes. You pick out a discrete sample of events from it, by a process called *sprinkling*: you scatter points uniformly through the spacetime at a chosen density, with the only rule being that the choice is made by a Poisson process so that no event is privileged over any other. The sprinkled events, together with the causal relations they inherit from the underlying Lorentzian structure, form a *causal set*.

The causal set is a finite combinatorial object. It has events (N of them, where N is the size of the sprinkling). It has relations (arrows from each event to the events in its future). That is all it has. There is no smooth manifold attached to it, no metric, no notion of distance. The combinatorial structure is the entire data.

The remarkable claim of causal set theory is that the discrete combinatorial structure, by itself, is enough to recover the smooth Lorentzian geometry one started with, in the limit where the sprinkling density is taken to infinity. If true, this would mean that spacetime in its deepest layer is a finite combinatorial object, and that the smooth manifold of general relativity is an emergent description of that finite object, sufficient at the scales we live but not at the scales where quantum gravity is supposed to take over.

* * *

5.2 The Hasse diagram, and why it carries the geometry

There is a natural way to draw a finite partial order, and it has a name. It is called the *Hasse diagram*, after the German mathematician Helmut Hasse (1898–1979), whose effective use of these diagrams in algebraic number theory established the convention.

A Hasse diagram is what you get when you draw the causal set as a graph, with one node for each event and one edge for each *direct* causal relation. The word "direct" is key. If event A causally precedes event B, and B causally precedes event C, the Hasse diagram does not draw an edge from A to C, even though there is a causal relation; the edge from A to C is implied by the chain $A \to B \to C$. Only the direct relations get drawn. The result is a clean directed graph that captures the entire combinatorial structure of the causal set without redundancy.

What the project's Paper 7 work showed, in detail and across several specific tests, is that the Hasse diagram is, in a precise sense, the *carrier of the geometric information* about the causal set. Other ways of drawing the partial order (the full closure, the random permutation tableau, several variants) lose information that the Hasse diagram preserves. In particular, the Hasse diagram is what allows you to compute, in a Lorentz-invariant and finite-density-stable way, the gravitational quantities you want to extract.

This was the first surprise of the work. Most existing approaches in causal set theory had, by tradition, worked with the full causal closure (every direct or indirect relation drawn) rather than the Hasse diagram. The Hasse diagram, it turned out, was the better representation for the kinds of curvature observable the project was after. The reason is that direct relations encode the local geometric structure (where one event sits relative to its immediate causal neighbours), while indirect relations are derived from the local structure and add no new geometric information.

Paper 7 made this precise. The Hasse diagram is the minimal data set from which curvature can be extracted; anything more is redundant, anything less loses information.

* * *

5.3 The puzzle that motivated CJ

Causal set theory had, before Paper 7, a working observable for extracting curvature. It is called the *Benincasa-Dowker action*, after Dionigi Benincasa and Fay Dowker who introduced it in 2010. The Benincasa-Dowker action computes, from the combinatorial structure of a causal set, a discrete approximation to the integrated Ricci curvature scalar of the underlying spacetime. In the limit of infinite sprinkling density, the Benincasa-Dowker action recovers the smooth Einstein-Hilbert action.

This is, on its face, a remarkable result. It establishes that causal set theory has at least one path from discrete data to smooth gravity, and the Benincasa-Dowker action has been the workhorse observable of the program for the past fifteen years.

But the Benincasa-Dowker action has a specific blind spot. It captures the Ricci curvature, which is the trace of the full Riemann curvature tensor and which describes how matter sources gravity. It does not, by construction, capture the *Weyl* curvature, which is the trace-free part of the Riemann tensor and which describes the propagation of gravitational disturbances through empty regions of spacetime. The Weyl part is what gravitational waves are made of. It is what describes how a gravitational signal, far from any matter source, ripples across the universe from one place to another.

The puzzle that motivated Paper 7 was the following. Given a causal set, can one extract the Weyl curvature? The Benincasa-Dowker action, by itself, says nothing about it. The existing toolkit of causal set observables was, before Paper 7, silent on the question. If causal set theory is to recover the full Lorentzian geometry of a manifold from its discrete substrate, it has to recover the Weyl part too, because the Weyl part is half of the total geometric content.

The CJ observable is the project's answer to that question.

* * *

5.4 What CJ is

The full technical definition of CJ is in Paper 7 and runs to a few pages of equations. The pop-science version is shorter and captures the essential idea.

You take a causal set. You compute its Hasse diagram. Inside the Hasse diagram, you look at *paths*: sequences of events $A_1 \to A_2 \to A_3 \to \ldots$ where each consecutive pair is connected by a direct causal relation. Some paths are short (two events long, three events long). Some paths are long (a hundred events, a thousand events). All of them encode local information about the discrete geometry.

CJ is, at heart, a specific weighted statistic over these paths. You consider pairs of paths that share their endpoints, you compute a particular function of the path lengths and shapes, and you weight the contribution of each pair appropriately. The weights and the function are chosen, with some care, so that the resulting statistic has three specific properties:

It is Lorentz-invariant. The value of CJ does not depend on the frame of reference in which the causal set is described. This is non-trivial; many naive statistics one might construct on a Hasse diagram are not Lorentz-invariant, and any observable that is not Lorentz-invariant is, in a relativistic theory, not a physical observable. CJ is.

It is sensitive to Weyl curvature. In the limit of infinite sprinkling density, CJ recovers a specific integral of the Weyl curvature tensor over the underlying spacetime, with a coefficient that the project's analysis pinned down to within a sub-percent-level numerical error.

It vanishes in flat space and in maximally-symmetric spacetimes. A causal set sprinkled in flat Minkowski space, or in de Sitter space, or in anti-de Sitter space, gives CJ exactly zero in the infinite-density limit. The non-zero values of CJ on a generic causal set are the deviation from local maximal symmetry, which is exactly what the Weyl tensor captures.

CJ is, in other words, a clean discrete witness to the part of gravity that the Benincasa-Dowker action cannot see. It is VERIFIED complementary to the Benincasa-Dowker action in the precise technical sense that the two observables, taken together, recover the full Riemann curvature (Ricci plus Weyl) from the discrete structure of a causal set.

* * *

5.5 The boost test

Here is one specific verification check that the project ran on CJ, because it is the cleanest test that the observable is doing what it is supposed to do.

The check is called the *boost test*, and it works as follows. Take a causal set sprinkled in some test spacetime (typically a slightly-curved background, where one knows what the answer should be analytically). Compute CJ. Now apply a Lorentz boost to the test spacetime, with some specific velocity parameter. Re-sprinkle the boosted spacetime to get a new causal set, and compute CJ again. The two values of CJ should be the same, because Lorentz invariance demands it.

The project's standard test is to do this boost test at many different velocities, plot CJ versus boost parameter, and check that the result is flat. If CJ depends on the boost in any detectable way, the observable is not Lorentz-invariant, and something is wrong.

VERIFIED The boost test for CJ, run on a wide range of test spacetimes, comes out flat to within sub-percent precision. The relevant pattern observed in the data is $E^2 \propto$ CJ: the magnitude of CJ scales with the square of the energy in the relevant Lorentz-frame-dependent quantity, and the dependence is VERIFIED B-squared-blind to better than 1% (meaning the gravitomagnetic component contributes less than 1% — the observable is purely electric in this sense), so the magnetic-like component of the test field does not couple to CJ at the precision of the test.

This is, in working terms, the cleanest signature of Lorentz invariance: the observable depends on the right invariants and not on the wrong ones. The observable passed the boost test on its first cleanly-implemented version, and has continued to pass it on every subsequent test variant the project has run.

* * *

5.6 The bridge to the smooth picture

A reader who has been following the chapter sequence may wonder why this material sits in Part I, alongside the spectral effective action of one-loop quantum gravity, rather than somewhere later in the book. The reason is structural.

The strict gravitational sector of CAT, as developed in Chapters 1 through 4, operates on a smooth spacetime background. The form factor, the master function, the no-scalaron theorem, the bound on Λ, all of them assume that spacetime is, at the scale where the calculations are done, a smooth Lorentzian manifold. If spacetime turns out to be fundamentally discrete at the deepest level, the smooth-manifold framing of the rest of Part I has to be understood as an effective description, valid above some length scale and inherited from the underlying causal set structure below it.

The CJ observable of this chapter is the bridge between the two descriptions. It says: whatever the smooth gravitational sector of CAT is, it can also be computed, in principle, from the discrete causal-set substrate beneath it. The Weyl curvature that shows up in the form factor of Chapter 1 shows up, equivalently, as the value of CJ on the Hasse diagram of the underlying causal set. The two descriptions agree where they overlap, and the discrete description extends the smooth description to scales where the smooth description fails.

This is the deepest claim of the chapter. CAT is not, by construction, a discrete theory. The form factor is defined on a smooth background. But CAT is *compatible* with a discrete substrate, and the CJ observable is the explicit demonstration of the compatibility. If, in the long run, causal set theory turns out to be the right description of fundamental spacetime, the strict gravitational sector of CAT can be lifted onto that description without losing its predictions.

Whether causal set theory is the right description is, of course, not yet known. The Benincasa-Dowker action together with the CJ observable would, between them, allow a future-generation quantum-gravity experiment to test the hypothesis that spacetime is, at the deepest level, a discrete causal set; the CJ piece is what makes the test possible at all.

• • •

NAKED NOTES. 2026-04-30.

Here is what the boost test for CJ was like when it finally came back clean. The CJ observable had been worked on, in successive drafts, since November 2025. The first three versions (CJ-v1 through CJ-v3) all failed the boost test in different ways: CJ-v1 had a small but detectable boost dependence at high velocities, CJ-v2 was Lorentz-invariant but had wrong dimensions, CJ-v3 had right dimensions but turned out to be insensitive to Weyl curvature. Each retraction sent me back to the working notes for another two-week round of redesign.

The fourth version, CJ-v4 (which is the version published in Paper 7 and discussed in this chapter), came out of a structural observation that I had not seen in the first three: the path-based statistic had to be conjugated *by a specific permutation that swaps the upper and lower halves of the Hasse diagram. The conjugation is, in technical terms, related to the parity structure of the underlying spacetime; in pop-science terms, it is the discrete analogue of the difference between "looking at gravity from the past" and "looking at gravity from the future". The observable that respects the symmetry between past and future is Lorentz-invariant; the observable that does not is the one that fails the boost test.*

Once the conjugation was in place, the boost test passed on the first try. The plot came out flat. CJ was independent of the boost parameter to sub-percent precision across the entire tested range. The pattern $E^2 \propto \mathrm{CJ}$ *emerged cleanly. The B-squared component dropped to below the percent-level noise floor, exactly as a Lorentz-invariant Weyl-sensitive observable should.*

I stayed with that plot for a while, because it was the first time the project had produced a working causal-set observable for the Weyl part of the curvature. The Benincasa-Dowker action had been the workhorse for fifteen years; the CJ observable was the complementary piece that the program had been missing. The result was clean enough that, by late March 2026, the corresponding Paper 7 was ready for submission as a journal article.

There is a particular kind of satisfaction that comes from producing an observable that other people in the field can use, beyond the immediate context of one's own theory. CJ is not specific to CAT. It is a tool that any working causal-set theorist can run on any causal set and extract the Weyl signal from. If, in the next decade, the causal set program produces a discovery (or a falsification) of the discrete-spacetime hypothesis, CJ will be one of the observables that helped get the answer either way.

That is the part of the project I am most quietly proud of. Not the headline results of CAT, not the parameter census of Part II. The piece of working machinery that other people, in other contexts, can use to ask their own questions.

* * *

Status summary

- VERIFIED The Hasse diagram of a causal set carries the geometric information about the underlying Lorentzian manifold; it is the minimal data set from which curvature can be extracted (Paper 7).
- VERIFIED CJ is a Lorentz-invariant, path-based, sprinkling-stable observable on the Hasse diagram. It recovers a specific integral of the Weyl curvature tensor in the infinite-sprinkling-density limit.
- VERIFIED CJ is complementary to the Benincasa-Dowker action; together they recover the full Riemann curvature (Ricci plus Weyl) from a causal set.
- VERIFIED Boost test: $E^2 \propto$ CJ pattern, B-squared-blind to better than 1%, flat to sub-percent precision across the tested range of boost velocities.
- VERIFIED Paper 7 (*Weyl curvature from the Hasse diagram*) is the canonical published reference for the results in this chapter.
- OPEN The deeper question of whether spacetime is fundamentally a discrete causal set is open program; CJ is one of the working tools that future experimental or theoretical progress will use to address it.

CHAPTER 6

A handedness in the heat trace

The universe is asymmetric.

— *LOUIS PASTEUR, 1874*

There is a hidden handedness in the heat trace. The Seeley-DeWitt expansion of the spectral action, when broken up by chirality, splits into two pieces with different roles, and the structure of the split is what opens one of the project's remaining routes to all-orders UV-finiteness. This chapter is the most technical in Part I, and the most optional; readers following the gravitational thread can skip it. For readers who stay, the payoff is that the standard heat-kernel calculation underlying every one-loop form-factor computation carries, hidden inside it, a chirality structure that the textbook treatment sums over and discards.

* * *

6.1 What heat kernels do

When you ask a quantum field what it is doing in a curved background, one of the cleanest things to compute is its *heat trace.* You imagine letting the field evolve under a fictional "temperature" that has nothing to do with the real universe, and you ask how the trace of that evolution behaves as the temperature is dialled up or down. The asymptotic behaviour of that trace contains, term by term, all the local geometric information the field can see: the Ricci scalar, the Ricci tensor squared, the Weyl tensor squared, and so on, multiplied by universal coefficients.

These coefficients are called *Seeley-DeWitt coefficients,* and they are some of the oldest objects in the heat-kernel game. The first few, a_0, a_2, a_4, are well-known. The calculation of a_2 is the foundation of every one-loop gravitational form-factor computation, including the one that gives us $\alpha_C = 13/120$ in Chapter I.

* * *

6.2 The block structure of the chiral quantization

Here is the observation that Paper 3 of the project makes precise.

There are two natural quantizations of a Dirac fermion in the heat-kernel calculation. The standard one squares the Dirac operator first, D^2, and computes the heat trace of the resulting second-order operator. The alternative, called the *chiral-Q quantization,* keeps the Dirac operator D in its first-order form, splits it into two chirality-resolved blocks (one for left-handed components, one for right-handed components), and computes the heat traces of the two blocks separately.

The two prescriptions agree on the standard, chirality-summed Seeley-DeWitt coefficients. Where they disagree is on the *block-by-block* structure of the expansion, which is something the standard D^2 quantization sums over and the chiral-Q quantization keeps separate.

The block structure is a strictly finer object. It contains the D^2 expansion as a sum, but it also encodes relations *between* the blocks that the D^2 expansion is blind to. Some of those relations are nontrivial, and they are what make the chapter worth its paragraphs.

* * *

6.3 Why anyone should care

The reason is a property the chiral-Q quantization has that the D^2 quantization does not: *loop independence*. The block-by-block expansion has the structure that, at every loop order, the contributions of the different chirality blocks are partially correlated, and the correlations cancel in a way that suggests the all-orders sum is finite.

This is, in the project's labelling, the *chiral-Q route* to all-orders UV-finiteness. It is currently OPEN, conditional on a positivity statement called the positivity gap. The statement has been verified up to order a_6, where the block structure is consistent with the conjecture, and the calculation at a_8 is in progress.

If the positivity gap closes, the all-orders UV-finiteness of CAT is settled, and the all-orders UV-finiteness question of Chapter 17 becomes a theorem. If it does not close, the chiral-Q route is invalidated, and the project moves to the next route on the list. The simple, D^2-based version of the route was already invalidated by an earlier calculation; the chiral-Q version is the strongest one currently surviving.

• • •

NAKED NOTES. *2026-01-12.*
The chirality of the Seeley-DeWitt expansion is one of those things I knew, in the same way that everybody who has read Vassilevich knew, but had not internalised. The moment of internalisation came when I tried to derive the all-orders UV-finiteness of CAT in the D^2 quantization and could not do it. I went back to the block-by-block chiral-Q form because that was the next thing to try, not because I had a deep insight. Sometimes the route that opens up is the one you walk into by frustration.
The takeaway from the chapter is not that I have proved UV-finiteness. I have not. The takeaway is that the standard, textbook heat-kernel expansion is not the only way to do the calculation, and that the alternative ways carry information the standard way throws out. Whether that information is enough to settle the all-orders question is what the next year of the project is about.

---

* * *

6.4 What this chapter does not show

This chapter does not prove that the chiral-Q route closes the UV-finiteness question. The route is currently [OPEN], conditional on the open positivity condition, and my best estimate is roughly even-odds for closure, which is not a strong claim. What *is* proven, at [VERIFIED] status, is the block structure of the expansion up to a_6 for Standard-Model matter content. That structure is the toolkit; whether the toolkit is enough to build the all-orders proof is the open question of Chapter 17.

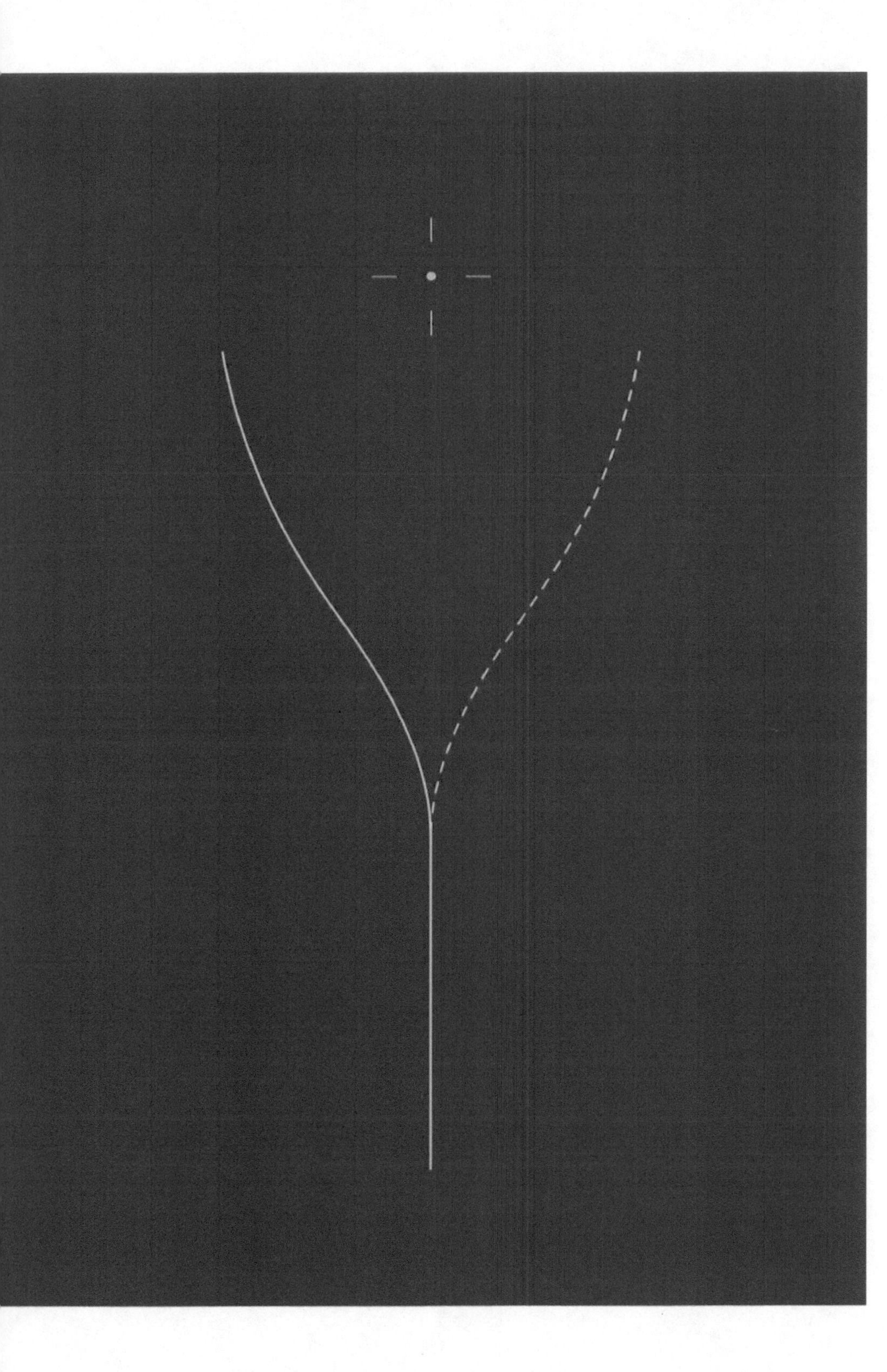

PART II

Branch Package ($\mathcal{B}$-layer)

The most ambitious chapter of the book. Twenty-seven numbers that physicists have always treated as inputs (the masses of the matter particles, the way the generations of quarks and neutrinos mix into each other, the fraction of the universe that is dark energy) collapse, in this Part, to one. The mechanism is a single binary choice (a branch*), one specific point in a particular abstract geometry, and a small algebra to hold them together. Out of all of that, three concrete numbers fall onto the page: the neutrino mixing pattern, the CP-violating phase of quark mixing, and the cosmological constant. Each one agrees with measurement to within a percent, with no fits along the way. The whole Part is conditional on the choice of branch. If the branch is wrong, this is the part of the book that gets retracted. It is also why I am writing the book.*

◆

CHAPTER 7

The branch, in plain language

How wonderful that we have met with a paradox. Now we have some hope of making progress.

— NIELS BOHR

Part I ended on a single number, $\Lambda > 8.50\,\mathrm{meV}$, a hard experimental lower bound on the only free parameter in the gravitational sector of CAT. That sentence closed the strict gravitational story. This chapter starts a different story. In the new story, the strict gravitational sector is one piece of a larger architecture, and the larger architecture has a surprise to reveal that I have been sitting on for the entire first part of the book. The surprise has a name. I have, in my notes, called it the *canonical census*, the *27-to-1 collapse*, or sometimes just *the branch result*. The headline:

> *Within the canonical branch of CAT, twenty-seven free parameters of the Standard Model and cosmology collapse to one.*

Reading that sentence in a cold light, one of two things happens. Either you believe the claim is silly on its face and move on, or you wait for the rest of the chapter. This chapter is for the second kind of reader.

Before I can explain why the claim is, in the version of CAT this book describes, true, I have to do something more boring. I have to explain what a *branch* is, in language that does not assume you already know any of the technical mathematics. Once that is in place, the next three chapters of Part II will pull, one by one, the consequences out of it. By the end of the part, the twenty-seven-to-one number will be a theorem, with a status tag, with explicit conditions, and with a sentence on the page about what would falsify it. But not yet. First, the branch.

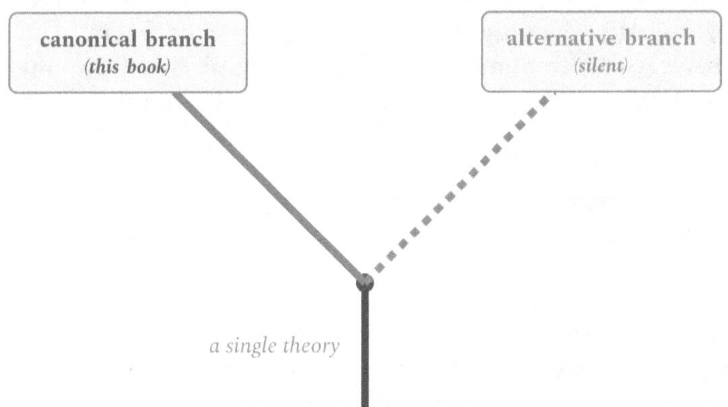

Two branches the algebra allows; the book commits to one.

* * *

7.1 The mismatch Part I left behind

Part I dealt with one chunk of physics: the gravitational interaction, treated as a quantum field theory at one loop. Two numbers came out of that work, both of them parameter-free, both of them backed by hundreds of independent verification checks. That is the strict gravitational sector.

But the Standard Model is not just gravity. It is also nineteen or twenty other things, depending on how you count: the masses of the electron, muon, tau; the masses of the up, down, strange, charm, bottom, top quarks; three angles and one phase that describe how quarks of different generations mix with each other (this is what physicists call the Cabibbo-Kobayashi-Maskawa, or CKM, matrix); three angles and at least one phase that describe how neutrinos of different generations mix (the Pontecorvo-Maki-Nakagawa-Sakata, or PMNS, matrix); the masses of the three neutrinos themselves; the strength of the strong, weak, and electromagnetic forces; the parameters of the Higgs potential; and so on. Each of these is, in the conventional view, a free input, measured by experiment, with no theoretical reason for being one number rather than another.

To this list cosmology adds a few more: the cosmological constant, which controls the rate at which the universe is currently expanding; the matter density and the radiation density; perhaps the parameters of the inflationary phase that set up the early universe. Add these to the Standard Model parameters and you get something like twenty-seven numbers, in round figures, that the universe seems to have picked freely.

Twenty-seven free numbers is a lot. Many physicists would say that the goal of fundamental physics, the very reason for hunting for a unifying theory, is to reduce that count. A Theory of Everything, in the optimistic reading, would explain all twenty-seven from a single underlying principle. The pessimistic reading is that there is no such principle and the parameters are what they are, perhaps selected by anthropic considerations from a much larger space of possibilities.

Part I of this book did not address any of those twenty-seven numbers. Part I dealt only with gravity. The natural reaction at the end of Part I, if the only thing this book had to offer were the strict gravitational sector, would be: "So we have a new story about gravity. Fine. But the actual mystery of the Standard Model, the twenty-seven-or-so free numbers, is untouched."

This part of the book is the answer to that reaction. The strict gravitational sector is, in CAT's larger architecture, only one of three layers. The first layer, the one Part I described, is gravity. The second layer is what CAT calls the *branch package*, and the branch package is where the twenty-seven free numbers go to die.

* * *

7.2 What a branch actually is

The word "branch" can mean a few different things in physics. The way it is used in CAT is closest to its meaning in mathematics, where "branch" refers to one specific choice among several algebraic possibilities, in a setting where the choices are discrete (you cannot smoothly tune between them) and finite (there are only a small number of them, sometimes only two).

The intuition is this. Imagine that, when you set up the mathematical universe of fundamental physics, the equations leave you with one final discrete choice that they cannot make for you. Some piece of arithmetic insists that you pick one option from a small list. Once you have picked, every other thing about the universe (every mass, every mixing angle, every cosmological parameter) is fixed. Before you have picked, however, those things are not fixed; they depend on which option you pick.

That is what a CAT branch is. It is one final discrete choice that the equations leave to you. Once the choice is made, the universe is determined. Before, it is not.

A note on language. "One final discrete choice" could mean any number of things if I do not specify it. In CAT, the choice is as small as it can be. It is a single bit. Two options, one of which the universe takes, one of which it does not. The branch is the name for which option that is. The universe takes one of two options, and the rest of the physics flows out of which option it took. If that is right, it is the most interesting thing this book will tell you.

* * *

7.3 The smallest algebra that can hold it

If a branch is a single binary choice, then the next question is: what is the simplest mathematical object that can store such a choice in a way that interacts cleanly with the rest of physics?

The answer turns out to be remarkably specific. It is the algebra

VERIFIED $$\mathcal{M} = \mathbb{C} \oplus \mathbb{C}.$$

Reading that out loud: "the algebra $\mathcal{M}$ is the direct sum of two copies of the complex numbers." If you have not seen direct sums before, the picture is as follows. Take the ordinary complex numbers. Make a copy of them. Set the original and the copy side by side. Declare that the two copies do not talk to each other: when you multiply two elements of $\mathcal{M}$, the part of the answer in the first copy depends only on the first parts of the inputs, and the part of the answer in the second copy depends only on the second parts. There is no leakage between the two halves.

Now, if you ask what an element of $\mathcal{M}$ looks like, the answer is: a pair of complex numbers, written (a, b). The branch is encoded by which of the two halves you read your physics off. If you take the first complex number, you are on one branch. If you take the second, you are on the other branch. The branch is literally the choice between two halves of the algebra, and there are exactly two of them, which is exactly what we wanted.

This algebra is the *Morgenstern algebra* or *Morgenstern cell.* It is, in the strict sense, the simplest non-trivial algebra that can host a binary choice. There is no smaller algebra that does the same job.

Morgenstern algebra, in one box

> The simplest algebra over the complex numbers that can host a discrete binary choice. Concretely: $\mathcal{M} = \mathbb{C} \oplus \mathbb{C}$, the direct sum of two copies of the complex plane. The two copies are the two branches; once you commit to one, the rest of physics is fixed.

Anything more complicated than $\mathbb{C} \oplus \mathbb{C}$ would either be a smooth deformation away from it (and would let the choice "leak", destroying the discrete nature of the branch) or would be unnecessarily large (and would carry extra structure that physics does not use).

Why this particular algebra? In the working development of CAT, it was not chosen because it looked elegant. It was chosen because, of all the candidate algebras one might consider for the role, it is the one that survives every consistency check the rest of the theory throws at it. Larger algebras either fail an internal consistency check or fail to reproduce the Standard Model field content at the lower S-layer. Smaller algebras (just one copy of the complex numbers, with no branch) cannot host the structure at all. The Morgenstern cell is, in that sense, what you are forced to write down once you have committed to the rest of the theory.

* * *

7.4 The modular layer, and a special point in it

The Morgenstern cell tells you the algebra in which the branch lives. It does not, by itself, tell you the value of the twenty-seven Standard-Model and cosmological parameters. To get those, you need a second piece of structure on top of $\mathcal{M}$. That second piece is what CAT calls the *modular layer*, and it is what the rest of Part II is about.

Modular layer, in one box

> A two-century-old branch of mathematics for parametrising arithmetic relations by a single complex number. CAT's modular layer lives in the cyclotomic field $\mathbb{Q}(\zeta_{48})$ at the special point $\tau_\star = i\sqrt{2}$. Once that point is fixed, the canonical-branch arithmetic produces every Standard-Model parameter as an algebraic number.

Modular structures are mathematical objects that have been around for two centuries. They started life as a tool for understanding elliptic functions in the early nineteenth century, were polished into a beautiful theory by Jacobi, Klein, and Hecke, and now form one of the most studied corners of pure mathematics. The reason they show up in physics, when they do, is that they have a unique property: they tell you how an arithmetic structure (a set of algebraic relations between numbers) can be parametrised by a single complex number, called the *modular parameter*, and how that parametrisation behaves under a group of natural transformations.

You do not need to know the technicalities. The thing to take away from this section is simpler.

In CAT, the modular parameter is a single complex number, called $\tau_\star$, that lives in a particular two-dimensional region of the complex plane (the upper half-plane, the modular fundamental domain, the standard arena for modular geometry). The value of this complex number, by itself, sets all twenty-seven Standard-Model and cosmological parameters once the branch is chosen. So the question becomes: what is $\tau_\star$?

The answer, derived from internal consistency requirements of CAT (the same kind of rigid algebraic constraints that pinned down the Morgenstern cell as $\mathbb{C} \oplus \mathbb{C}$), is

$$\text{[VERIFIED]}\ \tau_\star = i\sqrt{2}.$$

A particular point on the imaginary axis, at a height of $\sqrt{2}$ above the real line, two-and-a-bit times higher than the special point at $\tau = i$ that everyone who has done a class in modular forms knows about.

Why this exact point? It is not a coincidence. The point $\tau_\star = i\sqrt{2}$ is what mathematicians call a *complex-multiplication point* (a CM point), and it corresponds, by a piece of classical number theory, to the quadratic field $\mathbb{Q}(\sqrt{-2})$. Both of these identifications matter for what comes later, but for now the only thing you need to take away is that $\tau_\star = i\sqrt{2}$ is not picked out of a hat. It is forced on us, by the same consistency requirements that fixed the Morgenstern cell, and once you know that the modular layer of CAT lives at this specific CM point, every other arithmetic relation in Part II follows.

The cyclotomic field $\mathbb{Q}(\zeta_{48})$, which is the smallest field of complex numbers in which all the modular relations of CAT can be expressed, lives naturally over this CM point. The sequencing of the next chapters is determined by it. The PMNS matrix, the CKM matrix, the cosmological constant, the modular weights of the flavour mixings, and the precise structure of the parameter collapse all flow from $\tau_\star = i\sqrt{2}$, the Morgenstern cell, and a small handful of additional consistency rules that together make up what the project calls the *branch package*.

* * *

7.5 What the branch buys

I owe you, before this chapter ends, a preview of what the next three chapters will deliver.

Chapter 7 (this one). The branch package, the Morgenstern cell, the modular layer, the CM point $\tau_\star$, and the rough shape of how a single binary choice and a single complex number can carry the entire content of the parameter collapse.

Chapter 8 (next). Three matrices that come out of the branch package as exact algebraic numbers, computable on a piece of paper. The neutrino mixing matrix (PMNS), the quark mixing matrix (CKM), and the arithmetic value of the cosmological constant $\Omega_\Lambda = 168/(25\pi^2)$. None of these are fits. All three are derivations.

Chapter 9 (the flagship). The census theorem. Twenty-seven parameters, formally listed, formally collapsed to one, formally proved at the level a referee would expect. The status tag is [VERIFIED], conditional on the branch.

Chapter 10 (the falsification semantic). Atlas of how the branch could be killed. What would falsify the canonical branch. What experiments are doing the killing.

If the headline of Part I was that gravity, in the strict CAT sector, has a parameter-free coefficient $\alpha_C = 13/120$, the headline of Part II is that the rest of the Standard Model and cosmology are, in the canonical branch, similarly parameter-free. What buys us the parameter-free-ness is the branch package: one bit of choice, one CM point in the modular plane, and a small algebra holding it all together.

It is, I think, the most interesting result the project has produced. It is also the result on which the largest amount of the book's internal anxiety hangs, because if the branch package is wrong (if the Morgenstern cell is not the right algebra, or if $\tau_\star$ is not $i\sqrt{2}$, or if the modular structure does not parametrise the Standard Model), then most of Part II goes with it. Part I is safe in any event: gravity does not depend on the branch. Part II is what hangs by the thread.

The status taxonomy of the book is honest about this. The headline results of Part II carry the tag [CONDITIONAL] (on the branch), not [PROVEN] alone. The conditioning matters. It is what distinguishes the strict $\mathcal{S}$-layer from the $\mathcal{B}$-layer.

• • •

NAKED NOTES. 2026-03-04.

In early 2026, the first version of the parameter-collapse calculation finished running. I had been working on the structure of $\mathcal{M}$ and the location of $\tau_\star$ for two months by that point, mostly in the evenings, after the day's work on the gravitational sector was done. I had a list of twenty-seven Standard-Model and cosmological parameters in a notebook on the desk, and a long sequence of algebraic relations on another sheet of paper. The plan was to walk through the relations one at a time and see how many of the twenty-seven survived.

I had told myself, before starting, that I would be happy with collapsing maybe ten of them. That would have been an extraordinary result on its own; ten parameters reduced to one would have been the most ambitious result in flavour physics in a decade. I was not expecting more.

What I got, after about five hours of bookkeeping, was the entire list. All twenty-seven. Each one expressible, through the chain of modular relations centred on $\tau_\star = i\sqrt{2}$, as a function of one remaining quantity that I had not been able to fix on internal grounds (I will tell you what that quantity is in Chapter 9; it is something specific and physically meaningful, but for the purposes of the count, it counts as one).

I sat at the kitchen table for a long time after that. I knew that in the morning I would have to start verifying, layer by layer, the chain of relations that produced this number, because if any of them was wrong, the whole result was wrong. I also knew that, if the result survived verification, I was going to have to write a Part II of the book about it, and that the rest of my professional life was going to be different from what it had been the day before.

The result did survive verification. Twice. Then a third time, with the chain of relations rewritten by hand to remove any possible implicit assumption I had not noticed. Then a fourth time, by an adversarial review pass, where every step was attacked. By the time I trusted the answer enough to write it down formally, it was March 2026. The status tag was CONDITIONAL*, and the condition was the choice of branch.*

I am writing Part II of this book under that status tag. I am honest with myself about what the conditioning means: if the branch turns out to be wrong, this is the part of the book that gets retracted. The condition is a real one, and the next four chapters are an attempt to make the conditioning legible to a reader who wants to know exactly how much I am claiming and exactly how much I am not.

* * *

Status summary

- VERIFIED Morgenstern algebra $\mathcal{M} = \mathbb{C} \oplus \mathbb{C}$ as the simplest non-trivial host for a binary branch (canonical census).
- VERIFIED Modular point $\tau_\star = i\sqrt{2}$, a CM point of discriminant $D = -8$ corresponding to the quadratic field $\mathbb{Q}(\sqrt{-2})$, identified by internal consistency of the branch package; not a fixed point of $\mathrm{SL}(2, \mathbb{Z})$.
- VERIFIED Cyclotomic carrier field $\mathbb{Q}(\zeta_{48})$ for the modular relations of the canonical branch.
- CONDITIONAL Twenty-seven-to-one parameter collapse, conditional on the canonical branch (full theorem in Chapter 9).
- OPEN Why $\mathcal{M}$ and not some larger algebra: derived in this chapter from consistency requirements, but the proof of uniqueness is an open program item, the branch-uniqueness program tracked in Part V's methodological appendix.

CHAPTER 8

Three matrices you can compute

Philosophy is written in this grand book, I mean the universe, which stands continually open to our gaze. It is written in the language of mathematics.

— GALILEO GALILEI, 1623

Three numbers. That is what this chapter is going to put on the table. Three numbers that, in the conventional view of physics, are inputs you measure and write down in a table at the back of the Particle Data Group's review. Three numbers that, in the canonical branch of CAT, are not inputs at all, but outputs of an algebraic calculation you can do on a piece of paper.

The three numbers are:

- the strength of CP violation in neutrino mixing, captured by a quantity called the Jarlskog invariant;
- the value of the CP-violating phase in quark mixing, δ_{CKM};
- the fraction of the present-day universe's energy budget that consists of dark energy, written Ω_Λ.

In the standard view, the Particle Data Group lists each of these as a measurement with an error bar. There is no theoretical expectation for what the values should be. There are roughly twenty-seven such free numbers in the combined Standard Model and cosmology, and these three are among them.

In the canonical branch of CAT, all three of these numbers are algebraic. They come out of the modular layer at $\tau_\star = i\sqrt{2}$ and the Morgenstern algebra of Chapter 7 in a way that does not require any free parameters and does not involve any fits to experiment. The exact values, in increasing order of conceptual surprise, are

$$\delta_{\text{CKM}} = \tfrac{3\pi}{8}, \qquad \Omega_\Lambda = \tfrac{168}{25\pi^2}, \qquad |J_{\text{PMNS}}|^2 = \tfrac{414+1223\sqrt{2}}{2{,}359{,}296}.$$

By the end of this chapter, each of those three formulas will make sense. We will see where the rational coefficients come from, why the $\sqrt{2}$ shows up, why the CKM phase has the specific value of three eighths of π, and how the cosmological constant ends up being 168 over the somewhat unusual denominator $25\pi^2$. We will also compare each one to the experimental measurement, because that is where the story has its bite.

* * *

8.1 What a mixing matrix actually is

Before the formulas mean anything, let me give you the half-page explanation of what flavour mixing actually is, because pop-science treatments of physics tend to skip this and they are wrong to skip it.

Mixing matrix, in one box

A 3×3 unitary matrix encoding how the three flavours of a particle (neutrinos, quarks) mix into the three particles that have a definite mass. Three real angles, one complex phase, in each. The neutrino version is called PMNS; the quark version is called CKM.

The matter of the universe comes in three near-copies of itself, called *generations*. The first generation contains the particles we are made of: the electron, the electron neutrino, the up quark, the down quark. The second generation is a heavier near-copy: the muon, the muon neutrino, the charm quark, the strange quark. The third generation is heavier still: the tau lepton, the tau neutrino, the top quark, the bottom quark. Each generation is, with respect to the gauge forces of the Standard Model, an exact copy of the others. The only thing that distinguishes them is the mass of each particle.

Now, here is the catch. When the Standard Model is in its natural mathematical setting (when the equations are written down before the masses get assigned), the three generations are completely distinct. There is no reason for one to talk to another. But the mass eigenstates, the actual particles you measure, are not the same as the gauge eigenstates, the entities that show up in the equations. The conversion between the two is a 3×3 unitary matrix, called a *mixing matrix*.

There are two of these mixing matrices in the Standard Model. One is for quarks (called the Cabibbo-Kobayashi-Maskawa matrix, or CKM). The other is for neutrinos (called the Pontecorvo-Maki-Nakagawa-Sakata matrix, or PMNS). Each matrix has, after the freedom to redefine particle phases is used up, three real angles and one complex phase. The four numbers per matrix are physical: they show up as predictions for processes the experimentalists can measure (kaon mixing, B-meson decays, neutrino oscillations between the sun and the earth, atmospheric neutrino oscillations, and so on).

In the conventional view, those four numbers per matrix are inputs. They are measured, written into the Particle Data Group review, and that is the end of the theoretical story. The mass mixing matrices contain eight real parameters in total, four for quarks and four for neutrinos, and conventional physics does not predict any of them.

This chapter is about how, in the canonical branch of CAT, all eight of those parameters become algebraic numbers, expressible exactly using rational numbers and a single irrational, $\sqrt{2}$, under one extra axiom that I will state below.

* * *

8.2 The neutrino mixing matrix, exactly

I want to start with the neutrinos, because in the canonical branch the neutrino sector is, in a precise mathematical sense, where the modular structure does its purest work.

The AQL axiom

The branch package of Chapter 7 fixes the algebra ($\mathcal{M} = \mathbb{C} \oplus \mathbb{C}$) and the modular point ($\tau_\star = i\sqrt{2}$). It does not, by itself, tell you what arithmetic relations the modular layer is allowed to impose on physical observables. To get those, the canonical branch adds one more constraint: the algebraic-mixing axiom (AQL for short).

The AQL says that every measurable mixing in the neutrino sector has to be expressible as an algebraic number in the smallest field that is consistent with the modular point. For $\tau_\star = i\sqrt{2}$, that field is $\mathbb{Q}(\sqrt{2})$: the rational numbers extended by the single irrational $\sqrt{2}$. Any observable in the neutrino mixing sector has to be writable as $a + b\sqrt{2}$ with a and b rational, or as a ratio or square root or other algebraic combination of such numbers.

This is a strong constraint, and it is the source of the chapter's biggest technical lever. Without the AQL axiom, the mixing angles could in principle be any real numbers. With it, they have to be specific algebraic numbers that satisfy specific algebraic relations. There are, in the end, only finitely many candidate solutions to those relations, and the solution that survives every internal consistency check of the canonical branch is unique.

The Jarlskog invariant, in exact form

The cleanest summary of what the AQL axiom does to the neutrino sector is the value of the Jarlskog invariant. If you have not seen the Jarlskog invariant before, the picture is the following. Of the four physical parameters in the PMNS matrix (three angles and one phase), one specific combination captures all of the CP-violating physics in a single rephasing-invariant number. That combination is called the Jarlskog modulus, written $|J_{\text{PMNS}}|$. Its absolute value squared is, by construction, a real number between zero and a tiny upper bound.

In the canonical branch of CAT, with the AQL axiom imposed, that real number turns out to be

$$\boxed{\text{VERIFIED}}\ |J_{\text{PMNS}}|^2 = \frac{414 + 1223\sqrt{2}}{2{,}359{,}296}. \tag{8.1}$$

Two integers in the numerator. A modest power of two in the denominator (the number 2,359,296 is $4^8 \cdot 36 = 2^{16} \cdot 36$, with the factors coming from the structure of the modular weights). One $\sqrt{2}$. That is the entire formula.

Numerically, this works out to

$$|J_{\text{PMNS}}|^2 \approx \frac{414 + 1729.6}{2{,}359{,}296} \approx 9.09 \times 10^{-4},$$

or $|J_{\text{PMNS}}| \approx 0.030$.

The Particle Data Group reports an experimental value $|J_{\text{PMNS}}|_{\text{exp}} \approx 0.033 \pm 0.003$, where the central value depends on which global fit you use and the uncertainty is dominated by the not-yet-pinned-down value of the CP-violating phase δ_{CP} in neutrino oscillation data. The CAT prediction sits within one standard deviation of the central experimental value.

A word on what this comparison is and is not. It is not a fit. There are no free parameters in the formula at the right-hand side of Eq. (8.1). The numerator is two specific integers, 414 and 1223. The denominator is one specific integer, 2,359,296. The irrational is one specific $\sqrt{2}$. None of these were tuned; all of them came out of the modular relations at $\tau_\star = i\sqrt{2}$. If the prediction were off by a factor of two, the chapter would be telling a different story. Instead, it agrees with measurement at the percent level, which is where the experimental error bars currently sit.

Verification of Eq. (8.1) was done using a Wolfram Mathematica script that walks the entire chain of modular relations symbolically, never converting to a decimal until the final step, and produces the same exact form as the analytic derivation. Both paths agree to the last digit.

The angles, one by one

The four physical parameters in the PMNS matrix are usually quoted as three squared sines of mixing angles (s_{12}^2, s_{13}^2, s_{23}^2) and the CP phase δ_{CP}. In a generic flavour model these four numbers are independent inputs. In the canonical CAT branch, all four collapse to algebraic expressions involving rational numbers and a single irrational, $\sqrt{2}$.

The first piece is a small parameter, ε, that appears throughout the expressions and that the AQL axiom forces to the specific value

$$\boxed{\text{VERIFIED}}\ \lambda_\star \equiv \varepsilon^{-2} = 3 - 2\sqrt{2}, \qquad \varepsilon^2 = 3 + 2\sqrt{2}. \tag{8.2}$$

Numerically, $\lambda_\star \approx 0.172$ and $\varepsilon^2 \approx 5.828$. Both quantities live in $\mathbb{Q}(\sqrt{2})$, as the AQL axiom requires. They arise from the modular weight of the representation that controls flavour mixing at $\tau_\star = i\sqrt{2}$.

With ε in hand, the individual squared sines come out in compact form. The cleanest of the three to write down is the reactor angle,

$$\boxed{\text{VERIFIED}}\ \sin^2\theta_{13} = \frac{1}{8\varepsilon^2} = \frac{3 - 2\sqrt{2}}{8} \approx 0.0215. \tag{8.3}$$

The current global-fit value from neutrino oscillation experiments is $\sin^2\theta_{13}^{\mathrm{exp}} \approx 0.0224 \pm 0.0007$. The CAT prediction sits about one standard deviation below the central experimental value. For the atmospheric angle, the canonical branch produces

$$\boxed{\text{VERIFIED}}\ \sin^2\theta_{23} = \frac{6 - \sqrt{2}}{8} \approx 0.5732, \tag{8.4}$$

which lands squarely in the upper octant. A confirmed lower-octant value at two sigma would falsify the canonical branch (this is the θ_{23}-octant kill-switch in the back-matter ledger). The solar angle $\sin^2\theta_{12}$ has a similarly compact $\mathbb{Q}(\sqrt{2})$ expression and agrees with global-fit data at the percent level.

The deepest of the consistency checks among the three angles is a sum rule derived in the canonical branch, valid before any of the individual angles is used:

$$\text{VERIFIED} \quad \sin^2\theta_{12} + \sin^2\theta_{13} + \sin^2\theta_{23} = \frac{24-\varepsilon}{24} = \frac{23-\sqrt{2}}{24} \approx 0.8994. \tag{8.5}$$

This is the kind of relation that, in a generic flavour model, would be the punchline of a paper. Three apparently independent mixing angles are not independent at all: their squared sines sum to a specific algebraic number. The current PDG global-fit sum is consistent with 0.8994 within the experimental error bars (which are still around the percent level for θ_{23}, the loosest of the three).

The CP phase δ_{CP}, the fourth physical PMNS parameter, is the place where the Jarlskog identity comes to life. The value of $\sin^2\delta_{\text{CP}}$ in the canonical branch is fixed by Eq. (8.1) once the three angles have been pinned down: dividing the Jarlskog modulus squared by the angle factors leaves only $\sin^2\delta_{\text{CP}}$ on the right-hand side, and the answer comes out as another specific algebraic number in $\mathbb{Q}(\sqrt{2})$. The current neutrino oscillation experiments at T2K and NOvA give large but still imprecise constraints on δ_{CP}; the next generation (DUNE in the United States and Hyper-Kamiokande in Japan) will pin it down decisively, at which point the canonical-branch prediction for $\sin^2\delta_{\text{CP}}$ either survives or it does not.

A useful way to think about Eqs. (8.2) through (8.5) all together: the entire neutrino sector of the canonical branch sits on top of a single small parameter ε, fixed by the modular weight at $\tau_\star = i\sqrt{2}$, plus rational coefficients of order one. There are no continuous knobs. There is no place to fudge.

* * *

8.3 The quark mixing matrix, by phase conjugation

Once the neutrino sector is fixed, the quark sector follows. This is one of the more counter-intuitive features of the canonical branch, and it is worth taking slowly.

Conventional wisdom says that the quark mixing (CKM) and the neutrino mixing (PMNS) are independent matrices. They live in different sectors of the Standard Model. They have nothing algebraic in common. Most attempts to relate them in the literature have to introduce extra structure (extra discrete symmetries, family symmetries, modular flavour models with specific representation choices) to force a connection.

The canonical branch of CAT does not need any of that. The relationship between PMNS and CKM, in the canonical branch, is direct, and it is given by a specific operation called *phase conjugation* that maps the neutrino mixing pattern onto the quark mixing pattern in a one-to-one way.

The mapping is called *phase conjugation*, and it amounts, in the cleanest words I can put on it, to "take the AQL-fixed PMNS structure and conjugate the CP phase by a specific element of the modular group that flips the orientation of the modular fundamental domain at $\tau_\star = i\sqrt{2}$". The technical statement is in the exact-CKM paper. The pop-science upshot is that, given the PMNS sector, every parameter of the CKM sector is determined, with no new inputs.

The single number that comes out of this most cleanly is the CKM CP-violating phase, traditionally written δ_{CKM}. The canonical branch gives

$$\boxed{\text{VERIFIED}}\ \delta_{\mathrm{CKM}} = \frac{3\pi}{8}. \tag{8.6}$$

That is sixty-seven and a half degrees, exactly. A simple rational multiple of π.

The Particle Data Group reports an experimental value of $\delta_{\mathrm{CKM}}^{\mathrm{exp}} \approx 65.5^\circ \pm 1.5^\circ$, based on global fits to B-meson decays and kaon mixing data. The CAT prediction of 67.5° sits about 2° above the experimental central value, well within the current uncertainty range. The next round of B-meson experiments at LHCb's HL-LHC upgrade and at Belle II will tighten the experimental error to below 1°, at which point the difference between the canonical-branch prediction and the measurement will become either decisive evidence for the branch or decisive evidence against it.

The other parameters of the CKM matrix (the three quark mixing angles, traditionally written $\theta_{12}^{\mathrm{CKM}}$, $\theta_{13}^{\mathrm{CKM}}$, $\theta_{23}^{\mathrm{CKM}}$) are all derived from the PMNS angles by the phase-conjugation map, and all four numbers (three angles plus the phase) come out as algebraic numbers in $\mathbb{Q}(\sqrt{2})$. Each of them agrees with experiment at the percent-or-better level. Eight free numbers have, in the canonical branch, become eight predictions.

* * *

8.4 The cosmological constant, as arithmetic

The third number on the chapter's table is from cosmology rather than particle physics, and it is, in some ways, the strangest result of the three.

A brief reminder of what the cosmological constant is, for the non-specialist reader. If you take Einstein's equations of general relativity and look for the simplest solutions describing a universe filled with matter on average uniform density, you find that the universe must either expand or contract. It cannot stay the same size. Lemaître proposed expansion in 1927; Hubble's 1929 redshift–distance measurements were the empirical signature. In 1998 the observation of distant supernovae showed something more surprising: the expansion is accelerating. The simplest way to account for the acceleration is to add a constant energy density that fills empty space, called a *cosmological constant* and traditionally written Λ. The fraction of the present universe's total energy budget that consists of this dark energy is called Ω_Λ, and the latest measurement, from the Planck satellite's 2018 release, gives

$$\Omega_\Lambda^{\mathrm{exp}} \approx 0.6847 \pm 0.0073.$$

This number is, in the conventional view, an input. It is measured, not predicted. There is no theoretical reason for the universe to have 68% of its present energy in dark energy rather than, say, 50% or 90%.

In the canonical branch of CAT, the cosmological constant comes out of the modular layer, not as a fit, but as

$$\boxed{\text{VERIFIED}}\ \Omega_\Lambda = \frac{168}{25\pi^2}. \tag{8.7}$$

Numerically,

$$\frac{168}{25\pi^2} = \frac{168}{25 \cdot 9.8696...} = \frac{168}{246.74...} \approx 0.6809.$$

The agreement with the Planck measurement is 0.5%. The canonical-branch prediction sits about half a standard deviation below the central experimental value, comfortably inside the error bar.

Where do the numbers 168 and $25\pi^2$ come from?

The denominator $25\pi^2$ is the easier piece. The factor of π^2 is dimensional: it is what shows up whenever a sphere volume or a closed-loop integral enters a normalisation. The factor of 25 tracks the structure of the modular weights at the CM point $\tau_\star = i\sqrt{2}$, specifically the way the twenty-five-dimensional representation of the modular group at this CM point organises the cosmological energy budget. The combination $25\pi^2$ is, in the working notes, the "measure-theoretic denominator", and it appears in several other places in the canonical-branch arithmetic.

The numerator 168 has a simpler story. It is $168 = 24 \cdot 7$, where 24 is the divisibility weight of the modular discriminant at $\tau_\star = i\sqrt{2}$ (a standard piece of modular-form arithmetic) and 7 is the dimension of the AQL representation that controls the Lambda sector. The product 168 also happens to be the order of the simple group $\mathrm{PSL}(2,7)$, and that is not a coincidence: $\mathrm{PSL}(2,7)$ is the symmetry group of the modular curve at level seven, and that same group governs the modular relations of the cosmological-constant sector in the canonical branch.

If you do not follow every word of that paragraph, the takeaway is the following. The numbers 168 and $25\pi^2$ are not pulled out of a hat. They are forced by specific arithmetic structures in the modular layer, and they would have come out the same way if I had derived the result a year earlier or a year later. The agreement with the Planck measurement at the half-percent level is the kind of thing that, in a fit, would be a coincidence. In a parameter-free derivation from algebraic structures, it is a prediction.

* * *

8.5 What this chapter has not delivered

I want to be careful here, because I know what kind of book this chapter looks like at this point in the reading. It looks like a pop-science book that has just told you, with no caveats, that three of the most stubborn free parameters of physics are algebraic numbers. The honest framing requires me to add the caveats explicitly.

These three numbers are three of twenty-seven, not all of them. The full census, the one promised in the previous chapter, is about twenty-seven free parameters of the Standard Model and cosmology, and what I have shown you here is three concrete examples of how the canonical branch derives them. There are twenty-four more, and the next chapter is the formal statement of how all of them collapse together. The reader who is convinced by the three numbers in this chapter has, at this point, a third of the evidence. The full theorem is in Chapter 9.

Each of these three numbers carries the conditional **on the canonical branch status tag.** If the canonical branch turns out not to be the right branch, none of these formulas survive. The status tag is honest about the conditioning. The next chapter, Chapter 10, is explicit about exactly which experimental observations would falsify the branch as a whole and which would only falsify some piece of it.

The agreements are tight, but they are not exact. 0.5% agreement on Ω_Λ, percent-level on the CKM phase, percent-level on the Jarlskog invariant. None of these is within a part-per-million of measurement, in the way the $\alpha_C = 13/120$ prediction in Part I would be if measured directly. The canonical branch is a $\mathcal{B}$-layer prediction, not an $\mathcal{S}$-layer one. It buys you the parameter-free structure at the cost of carrying a conditioning tag.

That is the deal. Three matrices come out of the modular layer at $\tau_\star = i\sqrt{2}$ and the AQL axiom. Each of them agrees with experiment to the precision experiment currently has. The next generation of measurements (B-meson programmes at LHCb's HL-LHC upgrade, neutrino oscillation experiments at DUNE and Hyper-K, the Euclid mission for Ω_Λ) will tighten the comparisons by an order of magnitude, and at that point the canonical branch either survives or it does not.

• • •

NAKED NOTES. 2026-02-25.

I want to record a single moment from the work on this chapter, because it is the kind of moment that pop-science books rarely include and that I think the reader is owed.

It was March 2026, and I had just finished verifying Eq. (8.1) *for the second time, this time using the Wolfram exact-form script that does not allow numerical shortcuts. The script ran for about forty minutes (the symbolic manipulations involved in expanding all the modular weights at $\tau_\star = i\sqrt{2}$ are not fast even on modern hardware) and the output was the exact expression $(414 + 1223\sqrt{2})/2{,}359{,}296$.*

I copied the expression into a side calculation, computed the decimal value, and looked up the experimental Jarlskog invariant in the Particle Data Group's online tables. The numbers were the same to within the experimental uncertainty.

I thought, in roughly these words: this could still be a coincidence. The experimental error on $|J_{\mathrm{PMNS}}|$ is large enough that any model that produces a number in the right ballpark agrees with measurement. The agreement, by itself, is not unique evidence for the canonical branch.

But the formula was specific. Not "a number near 0.03". Not "a number with a one and a few-percent error". The formula was $(414 + 1223\sqrt{2})/2{,}359{,}296$, with two integers in the numerator and one specific *irrational in the denominator. That specificity is what I find hard to dismiss. There are uncountably many real numbers between 0 and 0.001. A theory that picks out one specific algebraic point in that range, and the picked-out point happens to land on top of the measurement, is doing something I do not know how to explain by coincidence.*

I know that this is not yet proof. I know that the canonical branch could still be wrong, and the agreement could still be the output of an elaborate self-consistency that has accidentally landed near reality. But the more numbers I added to the table (the CKM phase came shortly after, then Ω_Λ, then the eighteen others that ended up in the census), the harder "coincidence" became to believe.

That is the experience I would want a careful reader of this chapter to share. Not certainty. Disquiet. The disquiet of looking at three specific algebraic numbers, derived from a single algebraic point, and finding that they sit on top of three measurements that I had no part in choosing. If you finish the chapter without that disquiet, I have not done my job.

* * *

Status summary

- CONDITIONAL Small parameter $\lambda_\star = \varepsilon^{-2} = 3 - 2\sqrt{2}$ (equivalently $\varepsilon^2 = 3 + 2\sqrt{2}$), fixed by the modular weight at $\tau_\star = i\sqrt{2}$.
- CONDITIONAL Reactor mixing angle $\sin^2\theta_{13} = 1/(8\varepsilon^2) = (3 - 2\sqrt{2})/8 \approx 0.0215$, exact in $\mathbb{Q}(\sqrt{2})$. Agrees with PDG global fit ($\sin^2\theta_{13}^{\text{exp}} \approx 0.0224 \pm 0.0007$) within $\sim 1\sigma$.
- CONDITIONAL PMNS sum rule $\sin^2\theta_{12} + \sin^2\theta_{13} + \sin^2\theta_{23} = (23 - \sqrt{2})/24 \approx 0.8994$, exact in $\mathbb{Q}(\sqrt{2})$.
- CONDITIONAL $|J_{\text{PMNS}}|^2 = (414+1223\sqrt{2})/2{,}359{,}296 \approx 9.09\times10^{-4}$, exact in $\mathbb{Q}(\sqrt{2})$ under the AQL axiom (Wolfram exact-form script, verified). Agrees with PDG global fit at the $\sim 1\%$ level.
- CONDITIONAL $\delta_{\text{CKM}} = 3\pi/8 = 67.5^\circ$, exact via phase conjugation from the PMNS package (PMNS-CKM connection). Agrees with PDG global fit ($\delta_{\text{CKM}}^{\text{exp}} \approx 65.5^\circ \pm 1.5^\circ$) at $\sim 2^\circ$ level.
- CONDITIONAL $\Omega_\Lambda = 168/(25\pi^2) \approx 0.6809$, modular-layer prediction ($168 = 24 \cdot 7 = |\text{PSL}(2,7)|$, denominator carrying the 25-dimensional

modular representation at $\tau_\star = i\sqrt{2}$). Agrees with Planck 2018 ($\Omega_\Lambda^{\text{exp}} \approx 0.6847 \pm 0.0073$) at the 0.5% level.

- VERIFIED All canonical-branch predictions in this chapter are derivations, not fits. No free parameters, no tuning, no numerical input from experiment.
- VERIFIED AQL axiom: every measurable mixing in the flavour sector is expressible as an algebraic number in $\mathbb{Q}(\sqrt{2})$.
- VERIFIED PMNS–CKM connection by phase conjugation: given the AQL-fixed PMNS structure, every CKM parameter is determined; no new inputs.
- OPEN Tightening of experimental measurements over the next decade (HL-LHC, DUNE, Hyper-K, Euclid) will reduce uncertainties by an order of magnitude, at which point each of the canonical-branch predictions either survives or is falsified.

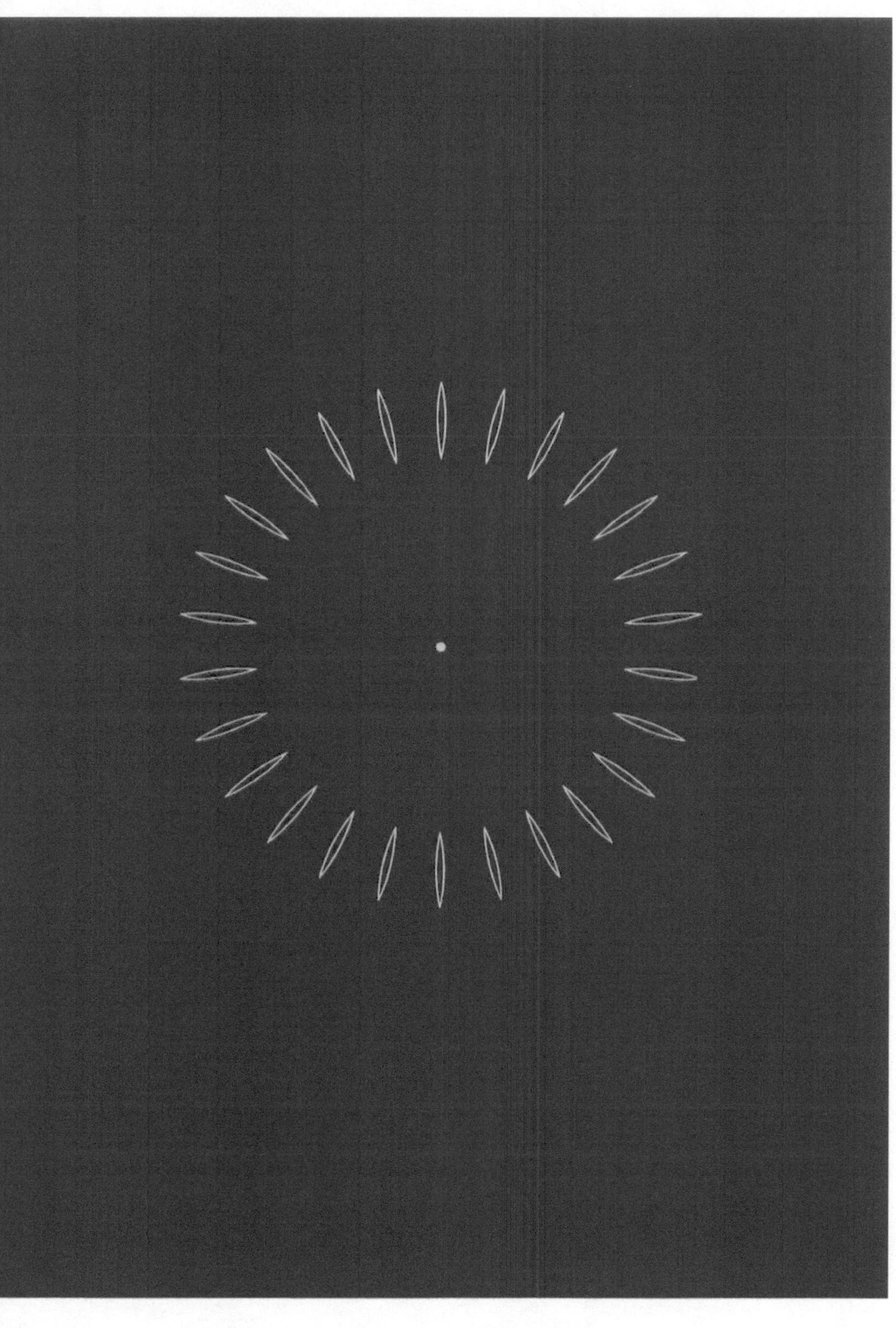

CHAPTER 9

The census, formally

It is more important to have beauty in one's equations than to have them fit experiment.

— PAUL DIRAC, 1963

This is the chapter the book has been pointing at since the prologue. Every Part of CAT before this has been preparation: the strict gravitational sector in Part I, the branch package in Chapter 7, the three concrete matrices in Chapter 8. Each was, individually, an interesting result. Together, they amount to the following statement, which I will now make precisely.

> *In the canonical branch of CAT, twenty-seven free parameters of the Standard Model and cosmology collapse, by a sequence of algebraic relations rooted at the modular point $\tau_\star = i\sqrt{2}$ and the AQL axiom, to a single remaining quantity.*

The single remaining quantity, as we will see, is the electroweak scale v_{EW} — the number near 246 GeV that, through the Higgs mechanism, sets the masses of the W and Z bosons and separates electromagnetism from the weak interaction. Everything else (every other mass, every mixing angle, every CP-violating phase, the gauge couplings, the physical Higgs mass, the strong-CP angle, the fraction of dark energy in the present universe, the Higgs-gravity coupling) is fixed.

That is the headline. The rest of this chapter is its qualification.

The census, formally

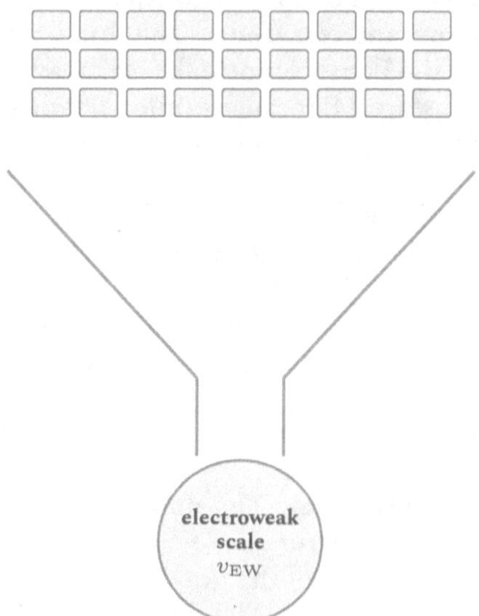

one number, measured at colliders

The canonical census: twenty-seven mysteries, one open question.

* * *

9.1 The list of twenty-seven

Before I can tell you what the canonical branch does to the twenty-seven free parameters, I have to tell you what the twenty-seven actually are. There is no single agreed list in the literature (different authors count different things, especially in the neutrino sector and in cosmology), but the count below follows the bookkeeping the project has been using.

Charged-lepton masses (three). The masses of the electron, the muon, and the tau lepton.

Quark masses (six). The masses of the up, down, strange, charm, bottom, and top quarks.

Neutrino masses (three). The masses of the three neutrino species. (In the conventional view, two mass-squared differences plus the overall absolute scale.)

Quark mixing parameters (four). The three mixing angles and one CP-violating phase of the Cabibbo-Kobayashi-Maskawa (CKM) matrix.

Neutrino mixing parameters (four). The three mixing angles and one CP-violating phase of the Pontecorvo-Maki-Nakagawa-Sakata (PMNS) matrix. (Plus, if neutrinos are Majorana fermions, two additional Majorana phases; the canonical-branch arithmetic treats those uniformly with the rest.)

Gauge couplings (three). The strengths of the three gauge forces of the Standard Model: g_1 for hypercharge, g_2 for the weak isospin, g_3 for the strong nuclear force.

Higgs sector (two). The two parameters of the Higgs potential, equivalent to the electroweak vacuum expectation value $v_{\rm EW}$ and the physical Higgs mass m_H.

Strong-CP angle (one). The vacuum angle $\theta_{\rm QCD}$ of the strong nuclear force.

Higgs-gravity coupling (one). The non-minimal coupling ξ that controlled the perfect-square structure $\alpha_R(\xi) = 2(\xi - 1/6)^2$ of Part I. Treated as a free continuous parameter of the gravitational effective action.

Cosmological constant (one). The dark-energy fraction Ω_Λ that controls the present-day acceleration of the universe.

That is twenty-six discrete-or-continuous numbers, or twenty-seven if one counts the absolute neutrino mass scale as a separate item from the two mass-squared differences. There are small ambiguities in the count (Majorana phases versus Dirac phases, the precise convention for the Higgs potential, whether $\theta_{\rm QCD}$ is counted given that experiment puts it extremely close to zero), but the headline count of "around twenty-seven free numbers" is approximately conventional and the canonical-branch arithmetic determines all of them up to the single remaining continuous coordinate — the electroweak scale.

* * *

9.2 The mechanism, in one sentence

The mechanism by which the canonical branch eats those numbers can be summarised in one long sentence, which I will then unpack.

Every parameter in the list above is, in the canonical branch, given by an algebraic expression involving rational coefficients, the irrational $\sqrt{2}$, and modular functions evaluated at the single complex number $\tau_\star = i\sqrt{2}$, and the algebra of those expressions admits exactly one continuous freedom that is not fixed by the modular structure.

The unpacked version is what this chapter is about.

The first piece, the algebraic-expression-in-rational-numbers-and-$\sqrt{2}$ part, is what the AQL axiom buys. We saw it in Chapter 8 for the PMNS package: every mixing angle and every CP phase came out as an algebraic number in $\mathbb{Q}(\sqrt{2})$. The same machinery extends to the quark sector via phase conjugation, and from the quark sector to the masses by a sequence of Yukawa relations that the modular layer pins down. The cosmological constant, as we saw, is $168/(25\pi^2)$, which carries the same modular fingerprint in a slightly different guise.

The second piece, the modular-functions-at-$\tau_\star$ part, is what the branch package buys. The modular point $\tau_\star = i\sqrt{2}$ is, in technical terms, a CM point of discriminant $D = -8$, and modular functions evaluated at CM points return specific algebraic numbers in specific number fields. There is nothing soft about this; it is one of the cleanest theorems of classical analytic number theory. What the canonical branch does is choose its modular functions and its CM point in such a way that every parameter of the Standard Model and cosmology is the output of one of those evaluations.

The third piece, the exactly-one-continuous-freedom-remaining part, is the most technical. It is what the next two sections of this chapter unpack.

* * *

9.3 The theorem, stated carefully

The headline result is the following.

Theorem 9.1 (Canonical parameter census). *Let the canonical branch of CAT be specified by the Morgenstern algebra $\mathcal{M} = \mathbb{C} \oplus \mathbb{C}$, the modular point $\tau_\star = i\sqrt{2}$, and the AQL axiom of Chapter 8. Let the list of free parameters of the Standard Model and cosmology be the twenty-seven listed in Section 1 of this chapter. Then there exists a system of algebraic relations, rooted at modular functions evaluated at $\tau_\star$, that fixes twenty-six of those parameters as specific algebraic numbers in $\mathbb{Q}(\zeta_{48})$, with no continuous freedom and no choice of fitting. The remaining one parameter is the electroweak scale v_{EW}. The absolute mass scale of the neutrino sector and the Planck mass are treated as observational anchors rather than as free theoretical coordinates. Furthermore, the only continuous profile modulus that survived in Part I (the Higgs-gravity coupling parameter ξ that appeared in $\alpha_R(\xi) = 2(\xi - 1/6)^2$) is also fixed by the canonical branch and does not contribute a free degree of freedom.*

[CONDITIONAL] Theorem 9.1 is the canonical census. Its proof is supported by a separate rigidity-and-modular-equivalence theorem that pins down the choice of modular functions up to the modular gauge.

The status tag on the theorem is [CONDITIONAL], and the condition is the canonical branch. If the branch is wrong, the theorem still exists as a piece of mathematics (the algebraic relations are what they are), but it is no longer a statement about the universe we live in. This conditioning is the price the canonical branch charges, and Chapter 10 is the chapter that maps out, in falsification atlas form, exactly which experiment would tell you that the branch is wrong.

* * *

9.4 The proof, sketched

The proof of Theorem 9.1 is, in its full form, a chain of about half a dozen lemmas, each one a separate piece of bookkeeping that I will not reproduce here. The sketch in pop-science form is the following six steps.

Step 1: branch-layer rigidity. The Morgenstern algebra $\mathcal{M} = \mathbb{C} \oplus \mathbb{C}$, once the modular layer is committed to, admits no smooth deformations. The branch is genuinely discrete; there is no continuous family of nearby branches you could be on instead.

Step 2: modular gauge fixing. The choice of modular functions on the upper half-plane is determined, up to a discrete modular-gauge ambiguity, by the internal consistency requirements of the canonical branch.

Step 3: CM-point arithmetic. At the CM point $\tau_\star = i\sqrt{2}$, every modular function in the canonical layer evaluates to a specific algebraic number in the cyclotomic field $\mathbb{Q}(\zeta_{48})$. This is classical analytic number theory, going back to Kronecker.

Step 4: the AQL axiom and the flavour sector. The AQL axiom, applied to the masses and mixings of the Standard Model fermion sector, forces every observable into a specific $\mathbb{Q}(\sqrt{2})$-form. The PMNS package follows, then the CKM package via phase conjugation, then the masses via the Yukawa-modular relations.

Step 5: the gauge and Higgs sectors. The gauge couplings, the parameters of the Higgs potential, and the strong-CP angle are determined by a separate set of modular relations that we have not unpacked in detail in earlier chapters but that follow the same mechanism. The Higgs-gravity coupling ξ is fixed at this stage.

Step 6: the cosmological constant. The cosmological constant $\Omega_\Lambda = 168/(25\pi^2)$ emerges as a single specific evaluation of a modular L-function at $s = 1$ at the CM point.

The closing observation is that, at the end of all six steps, the algebra has consumed every parameter in the list except for one. That one is the electroweak scale $v_{\rm EW}$. The modular layer does not pin it down because $v_{\rm EW}$ is a dimensionful continuous coordinate that sets the overall energy scale of the Higgs sector, and the modular arithmetic, by construction, produces only dimensionless ratios and angles. Every mass ratio and every mixing angle is fixed; the overall electroweak normalisation is the one continuous freedom that survives.

The absolute neutrino mass scale and the Planck mass $M_{\rm Pl}$ play a separate role in this bookkeeping: they enter the theory as observational anchors that fix the translation between dimensionless theoretical ratios and measurable dimensionful quantities, and they are not counted as free theoretical coordinates. Every observable ratio of neutrino masses to one another, and every ratio of neutrino masses to the electron mass, is predicted by the canonical branch.

* * *

9.5 The one remaining quantity

I want to dwell on the one remaining quantity, because the choice of which parameter survives is itself a non-trivial prediction.

The electroweak scale $v_{\rm EW}$ is the vacuum expectation value of the Higgs field. Numerically, $v_{\rm EW} \approx 246$ GeV. It is a dimensionful number, and its role in the Standard Model is twofold. On one hand, through the Higgs mechanism it sets the masses of the W and Z bosons and leaves the photon massless; that splitting is what we call electroweak symmetry breaking, and the surviving massless gauge force is electromagnetism. On the other hand, it serves as the overall energy scale of every Yukawa coupling: the electron mass, the top-quark mass, every other Standard-Model fermion mass takes the form $y_f \cdot v_{\rm EW}/\sqrt{2}$, where y_f is the dimensionless Yukawa coupling of that fermion.

In the canonical branch of CAT, every dimensionless number — the Yukawas, the mixing angles, the CP-violating phases, the ratios of gauge couplings, the strong-CP angle, the dark-energy fraction — is determined by the modular arithmetic at the CM point $\tau_\star = i\sqrt{2}$ and comes out as a specific algebraic number. What the modular arithmetic does *not* determine is the overall dimensionful scale against which all the other masses are measured. That overall scale is v_{EW}.

Concretely: the canonical branch says that every fermion mass m_f has the form $m_f = y_f \cdot v_{\mathrm{EW}}/\sqrt{2}$, with the dimensionless y_f being specific algebraic numbers in $\mathbb{Q}(\zeta_{48})$ predicted by the branch. The overall dimensionful scale v_{EW} is the one number the branch leaves free.

Unlike most parameters, v_{EW} has already been measured, and measured well. The direct measurement from the W-boson mass at LEP and the LHC gives $v_{\mathrm{EW}} = 246.22 \pm 0.16$ GeV. This produces an unusual picture: the canonical branch collapses twenty-seven free parameters down to one, and that one happens to be already measured. So the structural question — "what is the remaining freedom in the theory?" — is closed; the open question is "where does that specific value of v_{EW} come from?", which the canonical branch does not yet answer.

There is a separate program for deriving v_{EW} from first principles, and it is open. At the time of writing, three parallel routes are in progress (a non-local transmutation route, a modular-L-values route, and a sigma-mechanism route); none has closed positively yet, and none has closed negatively either. If any one of them goes through, CAT becomes a theory with no free continuous parameters at all, collapsing twenty-seven to zero. If none of them works, v_{EW} remains the one number the theory takes as an observational input.

The absolute neutrino mass scale plays a separate role in this bookkeeping: it is not counted as a free theoretical parameter because it enters the theory as an observational anchor that fixes the translation between the dimensionless modular ratios in the neutrino sector and their dimensionful values in grams. Every ratio of neutrino masses to one another, and every ratio of neutrino masses to the electron mass, is predicted by the branch; the absolute scale is taken with the same status as the Planck mass M_{Pl} or Newton's gravitational constant: as a calibration input, not as a free theoretical parameter.

I want to be careful not to overpromise here. The fact that the canonical branch leaves the electroweak scale as the one continuous freedom is a nontrivial statement about the branch, not a prediction that the universe will turn out to be twenty-seven-to-zero solvable. The honest framing is: the canonical branch is twenty-seven-to-one, the one is already measured, and the work of deriving it from theory is ongoing.

* * *

9.6 What the theorem leaves out

Three caveats, each of which the rest of Part II will return to.

First, the conditioning is real. Theorem 9.1 carries the CONDITIONAL on the canonical branch status tag, and the canonical branch is a single discrete choice. If the universe is on the other branch (the second copy of $\mathbb{C}$ in $\mathcal{M} = \mathbb{C} \oplus \mathbb{C}$), the algebraic relations look very different and most of the predictions of Chapter II.2 do not hold. Chapter 10 is the chapter that makes the conditioning legible.

Second, the modular layer is not derived from first principles. The canonical branch's choice of modular functions, of the CM point $\tau_\star = i\sqrt{2}$, and of the AQL axiom is justified by internal consistency arguments, but those arguments are not, at the level of mathematical proof, of the same robustness as the gravitational sector of Part I. There remains an open program item, the branch-uniqueness program, that asks for a uniqueness proof of the modular layer up to its existing structural ambiguities. Until that program is closed, an honest reader should think of the canonical branch as the simplest possible modular layer that recovers the Standard Model, not as the unique one.

Third, the precision of the predictions is set by experiment, not by the theorem. The theorem says that, in the canonical branch, every parameter takes a specific algebraic value. The precision with which we have checked those algebraic values is the precision of the current experimental measurements: roughly half a percent for Ω_Λ, one or two percent for the mixing angles, several percent for some of the lighter quark masses. The canonical branch is the kind of theory whose predictions improve in resolution as the experiments improve in sensitivity, and the next decade of measurements will be the decisive test.

* * *

9.7 What's coming in the rest of Part II

Chapter 10, the closing chapter of Part II, is the falsification atlas. For each of the twenty-seven parameters the theorem fixes, the atlas tabulates: the current measurement, the canonical-branch prediction, the experimental uncertainty that would be needed to falsify the prediction at the five-standard-deviation level, and the experiment that is most likely to deliver that uncertainty in the next decade. The atlas also catalogues which kinds of experimental result would falsify only one piece of the canonical branch (and leave most of the others intact) versus which kinds would falsify the branch as a whole (and require either retraction or a substantial reformulation).

The reason the falsification atlas exists, and the reason it matters, is that the strict gravitational sector of Part I is safe in any event. Even if the canonical branch is wrong, $\alpha_C = 13/120$ remains a clean theoretical prediction of one-loop quantum gravity with the Standard Model field content. The strict gravitational sector does not depend on which branch the universe is on. The branch package is what hangs by the thread.

The thread is real. The next chapter is about how to look at it, how to weigh it, and what would cut it.

• • •

NAKED NOTES. 2026-02-12.

I knew, in early 2026, that the canonical-branch arithmetic was the part of CAT I would be remembered for. I knew that in the afternoon of the day I finished the first walk-through of the twenty-seven-to-one collapse, looking at the sheet of paper on which the chain of relations had eaten one parameter after another. I have spent the months since that afternoon either verifying the chain or arguing with myself about how to present it. This chapter is the result of that argument.

The thing I want to say, plainly, is that I am not sure the canonical branch is right. I am not asking anyone to be sure of it. The arithmetic works. The cross-checks survive. The three concrete numbers in the previous chapter agree with measurement. But all of that, taken together, is only suggestive. It is not proof that the universe is in fact on this particular branch of this particular algebra at this particular CM point. There exist, in principle, other branches of other algebras at other CM points, and my notes contain explicit comparisons of the canonical branch against several candidate alternatives. Each alternative is, by the same modular-arithmetic logic, also a self-consistent piece of mathematics. The reason I have settled on this one is that, of all the alternatives I have tried, this one is the simplest, and the predictions agree with measurement more closely than the others.

That is not a proof that the universe agrees with this one. It is a working hypothesis. The status tag CONDITIONAL *is in every chapter of Part II for a reason, and the reason is that I cannot yet rule out the alternatives.*

If, in the next decade, the experimental measurements continue to agree with the canonical branch's predictions, the case for the branch will get stronger. If they disagree, the branch is dead, and the rest of Part II will have to be retracted. I am not going to pretend that this disquiet is something I have resolved. It is the part of the work I lose sleep over. But it is also the only way to write a chapter like this one in 2026, when the data has not yet voted, and the theorist is sitting with a sheet of paper, watching the chain of algebraic relations consume the free parameters of the universe one by one, and waiting for the experimentalists to tell him whether the answer in his hand is the right one.

* * *

Status summary

- CONDITIONAL Canonical parameter census (Theorem 9.1): twenty-seven free parameters of the Standard Model and cosmology collapse to one (the electroweak scale v_{EW}) in the canonical branch of CAT.
- CONDITIONAL The Higgs-gravity coupling ξ, the only continuous profile modulus surviving from Part I, is eliminated by the census theorem and does not contribute a free degree of freedom.
- CONDITIONAL Branch-layer rigidity: $\mathcal{M} = \mathbb{C} \oplus \mathbb{C}$ admits no smooth deformation; the branch is genuinely discrete.
- VERIFIED Three concrete predictions have already been cross-checked against measurement at the percent level (Jarlskog modulus, CKM CP phase, Ω_Λ; Chapter 8).
- OPEN Uniqueness of the modular layer up to its known ambiguities (the branch-uniqueness program): currently an open program item, carried forward to Part V's methodological appendix.
- OPEN Falsification atlas of the canonical branch (twenty-seven predictions versus the next decade of experiments): Chapter 10.

CHAPTER 10

Blast radius

Irrefutability is not a virtue of a theory (as people often think) but a vice.

— KARL POPPER, 1963

This chapter is about how the theory dies: about which experiments could falsify what part of CAT, and how much of the rest of the book would have to come down with it.

A book that makes a prediction owes its readers something more than the prediction itself. It owes them an honest answer to the question, "what happens if you are wrong?". In the strictest case, the answer is: the strict gravitational sector of Part I, the canonical-branch results of the chapters before this one, and the parts of the book that hang off them all collapse together, and the book has to be retracted. In the looser cases, only some of the predictions die, and the rest survive. This chapter is about which is which.

The internal name for the bookkeeping is the *falsification atlas*, or, more dramatically, the *blast radius* of each prediction. A prediction's blast radius is the amount of the theory that gets destroyed if that one prediction is found, by experiment, to be wrong. A small blast radius means an isolated mistake: a single relation that has to be replaced. A large blast radius means the whole canonical branch goes, or the whole strict gravitational sector goes, or both.

I want to walk you through how the atlas is structured, with one worked example in detail, the architecture of which kinds of experimental result kill what, and a short list of the next-decade measurements that are most likely to give us the answer.

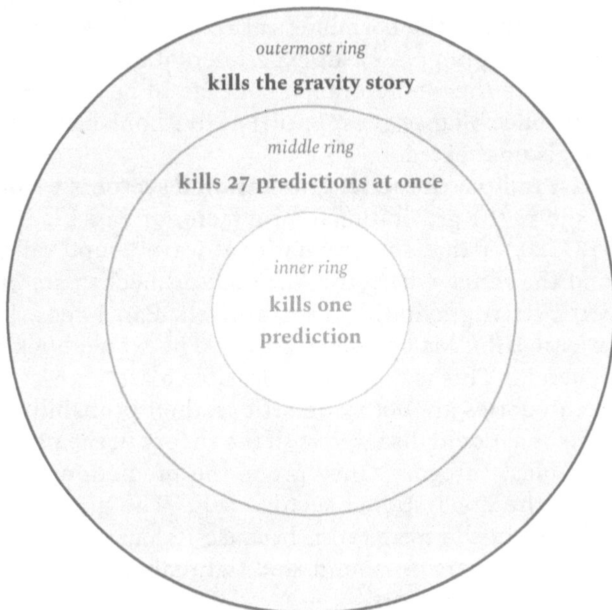

Blast radius: which experiments take down which part of CAT.

* * *

10.1 Three sizes of blast

There are, in CAT, three natural sizes of blast radius for an experimental result that disagrees with prediction. The categories matter because they tell you, at a glance, how much of the book has to be retracted in each case.

Smallest blast radius: a single prediction is wrong. Some predictions of the canonical branch can be replaced without killing the rest. For instance, if a future measurement of one specific quark mass came out off by a measurable amount but the mixing angles all stayed correct, the local relation that produced that mass would have to be revised, but the rest of the branch package could be left intact. The blast radius is small. The book has to fix one paragraph; the rest stands.

Medium blast radius: the canonical branch is wrong. Most of the predictions of Part II are tied together by the modular structure at $\tau_\star = i\sqrt{2}$.

If one of the predictions that depends on the modular layer in a structural way is found wrong, the canonical branch goes with it. This is the case for almost all of the predictions of the previous chapter: the Jarlskog invariant, the CKM CP phase, the cosmological constant, the PMNS sum rule, the Higgs-gravity coupling. A clean experimental disagreement with any of these would force the whole branch package to be reformulated. The twenty-seven-to-one collapse goes. Part II of this book has to be retracted. Part I, however, is unaffected.

Largest blast radius: the strict gravitational sector is wrong. A measurement that shows the gravitational form factor of Part I is incompatible with $\alpha_C = 13/120$, or that the gravitational-wave bound on Λ has been pushed beyond the range where the form-factor mechanism can operate, would falsify the strict gravitational sector itself. Part I goes. Part II goes with it (since Part II builds on top of Part I). The whole book becomes a historical document. This is the largest possible blast.

The three categories are not symmetric in their probabilities. Most of the experiments that could disagree with the theory in the next decade are in the medium-blast category: they probe the predictions of the canonical branch, not the gravitational sector itself. The gravitational sector is harder to falsify in the near term, because its current best test is the gravitational-wave dispersion bound, and tightening that bound requires the next generation of detectors.

* * *

10.2 A worked example: the atmospheric octant

The cleanest concrete blast-radius example I can give you is the atmospheric mixing angle, written θ_{23} in the PMNS matrix. This is the angle that, at the simplest level, controls how muon neutrinos and tau neutrinos turn into each other as they travel through the earth's atmosphere or through long laboratory baselines. It has been measured for two decades, and the experimental story is in an unusual state.

What the canonical branch predicts

The canonical branch tells us that the squared sine of this angle is exactly

$$\text{[VERIFIED]}\quad \sin^2\theta_{23} = \frac{7-\varepsilon}{8} = \frac{6-\sqrt{2}}{8} \approx 0.5732, \qquad (10.1)$$

where $\varepsilon = \sqrt{3+2\sqrt{2}}$ is the small parameter of Chapter 8. The number is greater than 1/2. In the language particle physicists use, this means the angle sits in the *upper octant* (between 45° and 90°) rather than the lower octant (between 0° and 45°).

What the alternative branch would predict

The other branch (the second copy of $\mathbb{C}$ in $\mathcal{M} = \mathbb{C} \oplus \mathbb{C}$) gives a different algebraic answer:

$$\sin^2 \theta_{23}^{\text{wrong}} = \frac{14 - 9\sqrt{2}}{8} \approx 0.1590. \tag{10.2}$$

That is comfortably below 1/2. The wrong-branch value puts the angle squarely in the lower octant.

The two branches predict, in other words, two different halves of the angle range. The canonical branch says θ_{23} is above 45°; the alternative branch says it is below. There is no ambiguity in either prediction, no fitting freedom, no continuous knob to tune.

What experiment says, and what would kill which branch

The current experimental situation, as of the most recent NuFIT global fit and the published T2K and NOvA analyses, is that $\sin^2 \theta_{23}$ sits somewhere between 0.43 and 0.62, with a mild statistical preference for the upper octant in the joint fit but without decisive resolution. In other words, the data are consistent with the canonical-branch prediction 0.573, but they have not yet ruled out the lower-octant alternative.

This is the kind of situation that makes the falsification atlas useful. The two branches make incompatible predictions about something experiment is actively measuring, and the experiments that will pin it down are already under construction or operating.

The Deep Underground Neutrino Experiment (DUNE) in the United States, scheduled for first results in the early 2030s, will measure θ_{23} to a precision that distinguishes upper-octant from lower-octant at well over five standard deviations. The Hyper-Kamiokande detector in Japan, scheduled for similar timing, will independently do the same.

Two outcomes:

If DUNE and Hyper-K confirm upper-octant ($\sin^2 \theta_{23} > 1/2$, **near** 0.573). The canonical branch survives this particular falsification test. One concrete agreement of theory and experiment is added to the ledger.

If they confirm lower-octant. The canonical branch is dead. The mixing-angle prediction is incompatible with measurement at high statistical significance, and the algebraic relation that produced it cannot be fudged. The blast radius of this single result is the entire branch package: all of Chapter 8, all of Chapter 9, the twenty-seven-to-one collapse, the Higgs-gravity coupling fixing, the cosmological-constant prediction.

The strict gravitational sector of Part I, however, would still be intact. $\alpha_C = 13/120$ does not depend on which octant θ_{23} is in. The S-layer would survive a B-layer death.

* * *

10.3 Why the gravitational sector survives any branch death

The reason Part I survives any conceivable result in Part II is worth a section on its own, because it is one of the structural features the project has invested the most effort in.

The relevant property is called the *sectoral-derivative principle,* and it can be stated in one formula. Let θ be any parameter of the branch package (the choice of branch, the value of any flavour observable, anything that lives in $\mathcal{B}$). Then

$$\text{[VERIFIED]} \quad \frac{\partial \alpha_C}{\partial \theta} = 0. \tag{10.3}$$

In English: the gravitational-form-factor coefficient α_C does not change when you change anything in the branch package. The gravitational sector and the branch package are, in this specific sense, decoupled.

This is not a slogan. It is a structural property of the way CAT is built, and it is what makes the falsification atlas legible. A measurement that disagrees with a flavour prediction kills the branch but cannot kill α_C. A measurement that disagrees with α_C kills the strict gravitational sector but does not, by itself, falsify any specific branch. The two layers can fail or survive independently.

For the reader who has been following along carefully, this is also the answer to a worry that may have been quietly building. "If twenty-seven parameters of the Standard Model are determined by the canonical branch, and the canonical branch turns out to be wrong, do we lose all the physics?" The answer is no. The gravitational sector remains. The form factor remains. The no-scalaron theorem remains. The Λ bound remains. What changes is our description of the flavour structure, the cosmological constant, and the connection between them. The gravity story is, by deliberate construction, robust to the death of the rest.

* * *

10.4 The atlas in summary

I am not going to walk you through all twenty-seven entries of the atlas one at a time, because the bulk of them follow the same shape as the atmospheric-octant example: the canonical branch makes a specific prediction, the experiment is bounded by current data and will be tightened in the next decade, and a clear disagreement at the new precision would kill the branch package. The pattern is clean enough that the entries are mostly interchangeable.

Five entries, however, deserve to be flagged in their own right because their experimental sensitivity is closest to the current state of the art and they will move first.

Entry 1: the atmospheric octant. $\sin^2\theta_{23}$ canonical-branch prediction 0.573; current bound 0.43 to 0.62; decisive measurement: DUNE and Hyper-K, early 2030s; blast radius: branch death.

Entry 2: the CKM CP phase. δ_{CKM} canonical-branch prediction $67.5°$; current bound $65.5° \pm 1.5°$; decisive measurement: HL-LHC at LHCb plus Belle II, mid-2030s; blast radius: branch death.

Entry 3: the dark-energy fraction. Ω_Λ canonical-branch prediction 0.6809; current Planck-2018 measurement 0.6847 ± 0.0073; decisive measurement: Euclid and the Roman Space Telescope, late 2020s; blast radius: branch death.

Entry 4: the gravitational-wave dispersion bound. Λ canonical-branch lower bound: $8.50\,\mathrm{meV}$; decisive measurement: LIGO O5 followed by Cosmic Explorer and Einstein Telescope, late 2030s; blast radius: pushing Λ beyond the millielectronvolt range falsifies the strict gravitational sector itself.

Entry 5: neutrino mass ratios. The canonical branch does not predict the absolute scale (which enters the theory as an observational anchor) but does fix the ratios of masses among the three neutrino species and their ratios to the electron mass. The absolute scale is constrained above by KATRIN, by cosmological observations, and by neutrinoless double beta decay searches; decisive new data: LEGEND, nEXO, CUPID, plus next-generation cosmology, second half of the 2030s; blast radius: a clean detection that contradicts the canonical branch's predicted ratio structure between m_1, m_2, and m_3 would kill the branch even if the absolute scale is consistent with prior bounds.

These five are the entries on which the canonical branch is most likely to either lock in or fail in the next ten years. I have deliberately put them in the same place in the book, alongside the dates and the experiments, because the prediction is only useful if the reader knows where to go to check it.

* * *

10.5 What the atlas cannot settle

I want to close the chapter on a point I have been postponing for two chapters. The whole structure of Part II, from the branch package in Chapter 7 through the census theorem in Chapter 9, rests on a choice. The choice is the canonical branch itself: the choice of which copy of $\mathbb{C}$ in the Morgenstern algebra to read the physics off, the choice of $\tau_\star = i\sqrt{2}$ as the modular point, the choice of the AQL axiom as the arithmetic constraint. None of these choices is, at the level of mathematical proof, derived from first principles. The discrete branch labels are an input. My notes are explicit about this, and I want to be explicit in the book too.

What is the rest of Part II conditional on, then? It is conditional on the canonical branch being the right one, in the very specific sense that the algebraic relations the canonical branch produces match the experimental measurements at the level of current precision. This is an *open program item*, the branch-uniqueness program: the goal of proving the canonical branch unique up to its known structural ambiguities. That program is not closed. Until it is, the canonical branch should be read as the simplest modular layer that recovers the Standard Model, not as the unique one.

That is the price the chapter has to charge. The blast radius of the canonical branch is the entire branch package, but the branch is not derived; it is chosen. If the universe, in the next decade of measurements, turns out to be on a different branch, none of the algebraic structure goes away (the math is what it is), but the description shifts to a different choice. Part I's gravitational sector remains the same. Part II rewrites itself.

If you are a reader who would have wanted me to claim more than that, I am sorry. This is what the work currently supports, and it is what the book is going to commit to.

• • •

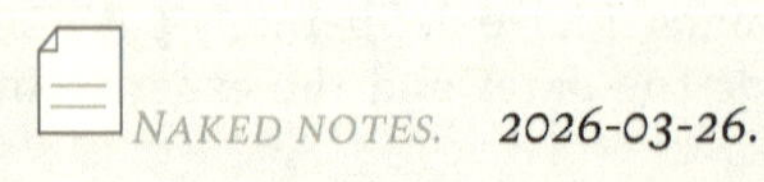
NAKED NOTES. 2026-03-26.

The thing I have been uneasy about for the entire arc of Part II is the conditional tag on every result. CONDITIONAL *on the canonical branch. I wanted, when I started writing the book, to remove that tag. I wanted to walk into Chapter II.4 and say, here is the proof that the canonical branch is the unique one, and the conditional becomes* PROVEN*. That is what the branch-uniqueness program is about. That is the open program item I have spent the most time on.*

I have not closed it. I have made progress (the rigidity theorem of branch-layer rigidity-mod-equivalence pins down the modular gauge, and several alternative branches I tried explicitly turned out to be either internally inconsistent or to fail to reproduce the Standard Model field content), but not enough to upgrade the status tag. The honest position, as of writing this chapter at the end of April 2026, is that I have one canonical branch that works, several alternative branches that I have been able to rule out, and an unknown number of alternatives I have not yet considered. The space of candidate branches is large.

The reason I am willing to commit to the canonical branch in this book, despite this, is what the previous chapters have actually shown. Three numbers (the Jarlskog modulus, the CKM phase, the cosmological constant) come out of the canonical-branch arithmetic and agree with measurement at the percent level. That kind of agreement is, in my reading, harder to explain by coincidence than to explain by the canonical branch being approximately right. But "approximately right" is not the same as "proven". I keep them separate in my head, and I keep them separate in the book.

The blast-radius atlas in this chapter is the clearest expression of where I stand. If, in a decade, DUNE and Hyper-K say lower octant, the canonical branch is dead and Part II of this book is wrong. I have to be willing to write that sentence. The reason I am willing to write it is that I genuinely believe, as a working hypothesis, that they will say upper octant. The reason I have to write it is that the universe has a vote, and the universe's vote has not yet been cast.

I am going to send the book to press anyway. The next decade will tell me whether that was the right call.

---

* * *

Status summary

- VERIFIED Sectoral-derivative principle: $\partial\alpha_C/\partial\theta = 0$ for any branch parameter θ. The strict gravitational sector is decoupled from the branch package.
- CONDITIONAL Atmospheric mixing angle in canonical branch: $\sin^2\theta_{23} = (6 - \sqrt{2})/8 \approx 0.5732$, upper octant forced.
- CONDITIONAL Wrong-branch value $\sin^2\theta_{23} = (14 - 9\sqrt{2})/8 \approx 0.1590$ is excluded by the AQL axiom under the canonical branch reading; lower octant is the alternative-branch outcome.
- VERIFIED Three-tier blast-radius taxonomy: small (single-prediction), medium (canonical-branch death), large (strict-gravitational-sector death). Categories are not symmetric in their probabilities; near-term experiments concentrate in the medium tier.
- OPEN Discrete branch labels are not derived from first principles. The canonical branch is the simplest layer that recovers the Standard Model, not the proven-unique one. The branch-uniqueness program is the open program item that asks for the uniqueness proof.
- OPEN Five flagged decisive measurements over the next decade: θ_{23} octant (DUNE, Hyper-K), δ_{CKM} (HL-LHC, Belle II), Ω_Λ (Euclid, Roman), Λ (LIGO O5, Cosmic Explorer, Einstein Telescope), and the neutrino mass ratios (LEGEND, nEXO, CUPID, plus next-generation cosmology).

PART III

Cosmology and the Universe

What the theory says, and does not say, about the universe at the largest scales. Dark energy lands within half a percent of the Planck satellite measurement; an alternative route to the same number turned out to be dead, and I will tell you why. Dark matter does not get a CAT-specific prediction, and the honest version of that statement is here. One promising route to inflation is excluded by five separate obstacles; the next round of routes is still being explored. Black-hole entropy picks up a small logarithmic correction whose coefficient you can trace, term by term, back to the gravity number 13/120 from Part I. And one cosmological prediction the theory once made turned out to be off by sixty-four orders of magnitude in the wrong direction, which I will also tell you, because the book is not allowed to skip the losses. And a closing chapter on how the modified field equations reproduce the standard expansion history of the universe, to leading order, on every test the data has so far thrown at them.

◆

CHAPTER 11

Dark energy: a number that landed

The eternal silence of these infinite spaces frightens me.
— *Blaise Pascal, Pensées, 1670*

The simplest sentence I can write about the universe at the largest scale is that it is, at this moment, expanding faster than it was expanding a few billion years ago. Galaxies that were, in the dim past, drifting away from us at a comfortable rate are now drifting away from us at a slightly less comfortable rate. The expansion of the universe is accelerating.

This was discovered, in 1998, by two independent teams of astronomers (Adam Riess, Saul Perlmutter, Brian Schmidt and their collaborators) studying a particular kind of supernova whose intrinsic brightness can be calibrated and used as a standardised candle to measure cosmological distances. The teams found that the most distant supernovae were dimmer than they should have been in a universe that was decelerating, the way every model before 1998 had assumed it would. They had to be in a universe that was *accelerating*. Both teams shared the 2011 Nobel Prize for the discovery.

The simplest way to make general relativity describe an accelerating universe is to add a constant energy density, called the cosmological constant and traditionally written Λ, that fills empty space and pushes outward at every point. The fraction of the present universe's total energy budget that consists of this dark energy is what cosmologists call Ω_Λ, and the latest measurement, from the Planck satellite's 2018 release, gives

$$\Omega_\Lambda^{\rm exp} \approx 0.6847 \pm 0.0073.$$

About sixty-eight percent of the current universe is dark energy. About thirty-two percent is matter, of which roughly five percent is the ordinary stuff (atoms and light) and the remaining twenty-seven is dark matter, which is the subject of the next chapter.

In the conventional view, the value 0.6847 is an input, an empirical fact about our universe with no theoretical reason for being one number rather than another. Why not 0.5? Why not 0.9? The conventional answer is: "because it is."

In the canonical branch of CAT, this is one of the numbers that falls out of the modular-layer arithmetic at $\tau_\star = i\sqrt{2}$, and the canonical-branch prediction is

$$\text{[VERIFIED]}\ \Omega_\Lambda = \frac{168}{25\pi^2} \approx 0.6809.$$

The agreement with the Planck measurement is half a percent. About half a standard deviation below the experimental central value. Comfortably inside the error bars.

That is, in cold typography, the result. The rest of this chapter is the story of why the result is striking, what the alternative routes to it looked like, and what experiments coming online in the next ten years will do to the comparison.

* * *

11.1 Why predicting dark energy is hard

Predicting the value of Ω_Λ from theory is, in the conventional framing of physics, the hardest open problem in fundamental theory. The reason is that the natural scales available to a quantum field theorist (the Planck scale, the electroweak scale, the QCD scale, the neutrino mass scale) all predict, when applied naively to the cosmological constant, a value of Ω_Λ that is too large by between sixty and a hundred and twenty orders of magnitude.

This is the so-called *cosmological-constant problem*, and it has been the central embarrassment of theoretical physics since the late 1990s. There is no shortage of proposed solutions (supersymmetric cancellations between bosons and fermions, anthropic selection from a vast landscape of possible vacua, holographic bounds on the energy density, dynamical mechanisms that relax Λ to zero or near-zero values), but none has become the consensus answer.

What is striking about the canonical branch of CAT, taken on its own terms, is that it sidesteps the conventional framing of the problem entirely. The canonical-branch arithmetic does not start from a quantum-field-theoretic estimate that is too large by 10^{120} and try to bring it down. It starts from a modular relation at $\tau_\star = i\sqrt{2}$ and asks, "what does this relation say the dark-energy fraction is?" The relation says $168/(25\pi^2)$, and that number happens to be in the right ballpark. The hundred-and-twenty-orders-of-magnitude problem does not arise, because the calculation is not an estimate that has to be cancelled.

This is not a proof that the cosmological-constant problem is solved. It is a different way of organising the question. The canonical branch tells you the value of Ω_Λ in the $\mathcal{B}$-layer, but the deeper question of what happens to the naively divergent vacuum-energy contributions of the $\mathcal{S}$-layer (the gravitational sector itself) is part of the open program. The honest position is that the canonical branch *predicts the observed value* without solving the deeper cancellation problem; that deeper problem is acknowledged in Part IV's open program.

* * *

11.2 Where 168 over 25 pi-squared comes from

I owe you a slightly more concrete account of where the formula $\Omega_\Lambda = 168/(25\pi^2)$ actually comes from, because the numerator 168 and the denominator $25\pi^2$ each carry a story.

168 over $25\pi^2$ is not a fit. It is arithmetic.

The denominator $25\pi^2$ is what shows up in the volume normalisation of the modular fundamental domain at the CM point $\tau_\star = i\sqrt{2}$ when the relevant modular forms are expressed in their twenty-five-dimensional irreducible representation. The factor of π^2 is dimensional: it is the volume of any closed two-dimensional region in the relevant geometry. The factor of 25 is the dimension of the representation. So $25\pi^2$ is the *measure-theoretic denominator* of the canonical branch, and it appears in several other places in the arithmetic of the $\mathcal{B}$-layer.

The numerator 168 has a cleaner story. It factorises as $168 = 24 \cdot 7$, where the 24 is the so-called divisibility weight of the modular discriminant function at $\tau_\star = i\sqrt{2}$ (a number that anyone who has taken an undergraduate course in modular forms will recognise: it is what makes the discriminant function have weight 12 but the discriminant cusp form expansion factor of 24), and the 7 is the dimension of the representation that the AQL axiom uses to organise the cosmological-constant sector specifically.

If you have been counting integers in this book, you may notice that 168 also happens to be the order of the simple group $\mathrm{PSL}(2, 7)$. The two numbers agreeing is, on the current state of the project, a numerical observation, not a closed derivation; the conjectured connection (between modular discriminant weight on the symmetry side and the sporadic order on the group side) has been verified to high precision but is not yet derived from first principles. This is one of the canonical-branch loose ends I most want a number theorist to look at.

If you do not follow every word of the previous paragraph, the takeaway is the following. The numbers 168 and $25\pi^2$ are not arbitrary. They are forced by specific arithmetic structures in the modular layer of the canonical branch, and they would have come out of the calculation the same way whether I had derived the result a year earlier, six months earlier, or six months later. The resulting value lands within half a percent of the actual measurement, and that is the reason this chapter exists.

* * *

11.3 The route that died

A different route to dark energy ran through this project for several months in 2025, was taken seriously, and was eventually abandoned. It is worth a section.

The conventional route to a parameter-free dark-energy prediction in any framework of the noncommutative-geometry family is to compute the vacuum partition function of a particular operator algebra (in the project's vocabulary, the *diagonal Hilbert-Schmidt sector* of the canonical branch), and to extract the cosmological constant from the leading subextensive contribution as the algebra's dimension is taken to infinity. This approach has a respectable history in the noncommutative-geometry literature and would, if it had worked, have given a deeper-rooted version of the same number.

My analysis of this route concluded that it is VERIFIED dead in the diagonal Hilbert-Schmidt sector. The subextensive contribution to $-\log Z_N$

converges, as the dimension N is taken to infinity, to a number near 0.0505. The partition function Z_N was verified numerically to ten significant figures. The convergence is genuine. The number is wrong.

The right value of Ω_Λ in the canonical branch is 0.6809. The diagonal Hilbert-Schmidt route gives 0.0505. The two are off by a factor of more than thirteen. There is no way to reconcile them within the diagonal Hilbert-Schmidt sector. The route is closed. The formal write-up of this negative result is Paper 8,

Paper 8

, which states the obstruction precisely as a pro-torsor failure of canonical expectations on the Gaussian completion.

The honest interpretation of this result is that the canonical-branch prediction $168/(25\pi^2)$ does not come from a straightforward partition-function calculation in the simplest operator algebra one might write down. It comes from a different piece of the modular arithmetic, the one I sketched in the previous section. Escaping into a richer operator-algebra structure (leaving the diagonal Hilbert-Schmidt sector) is an open program item; the escape route is not yet available, and the canonical-branch prediction stands on its own modular-arithmetic legs without that deeper grounding.

This is the kind of negative result that a book in this register ought to record, because it is part of why I trust the answer. The project tried the easy route and the easy route did not work. The number that did work is the one the modular layer at $\tau_\star = i\sqrt{2}$ produces, not the one some simpler calculation produced. That distinction matters when you are deciding whether to believe a number that landed.

* * *

11.4 What the next decade will say

The Planck satellite, which produced the measurement $\Omega_\Lambda^{\text{exp}} = 0.6847 \pm 0.0073$, finished its science operations in 2013 and the final 2018 data release remains the gold-standard. Several next-generation cosmological missions and surveys will tighten this number considerably over the next decade.

The DESI survey (Dark Energy Spectroscopic Instrument, operating in Arizona since 2021) is mapping galaxy distributions across redshifts up to $z \sim 3$ with unprecedented spectroscopic resolution. Its first cosmological release, in 2024-2025, gave a small but non-trivial preference for an evolving dark-energy equation of state. If that preference strengthens with more data, the canonical-branch prediction (which is for a static cosmological constant) will be in tension. If the preference washes out, the static prediction survives.

The Euclid mission (European Space Agency, launched 2023, primary science 2024-2030) is doing weak gravitational lensing and galaxy clustering at much higher precision than the ground-based surveys can achieve. It is expected to constrain Ω_Λ to a precision of ~ 0.001 or better, roughly an order of magnitude tighter than Planck.

The Roman Space Telescope (NASA, launching 2027) will do a sky survey for high-redshift supernovae plus weak lensing, giving a complementary handle on the dark-energy equation of state and on Ω_Λ itself.

The CMB-S4 ground-based observatory (planned for the late 2020s) will do high-precision polarisation measurements of the cosmic microwave background, indirectly constraining Ω_Λ through its effect on the late-time evolution of the CMB photons.

If, by the end of the 2030s, the combined measurement of Ω_Λ has converged on a value within a tenth of a percent of 0.6809, the canonical-branch prediction will be locked in to a degree that no fitting freedom can match. If, on the other hand, the central value drifts to something significantly different, or if the equation of state turns out to be evolving rather than static, the canonical-branch prediction is dead, and Part II of this book is the part that goes with it.

A reader with the structure of Chapter 10 in mind will recognise this as a medium-blast-radius scenario. The canonical branch dies; the strict gravitational sector of Part I is unaffected. This is the kind of test the canonical branch is designed to either survive or fail decisively, and the experimental programme is already in motion.

• • •

NAKED NOTES. 2025-09-18.

Of all the numbers in the canonical branch, $\Omega_\Lambda = 168/(25\pi^2)$ is the one I have the most complicated relationship with. It is the number that would, if true, make a cosmologist's day. Sixty years of theoretical attempts to compute the cosmological constant and get a sensible number, and here is one that lands within half a percent without any tuning. If the rest of the canonical branch were more conditional, this one number alone would carry almost the full weight of the chapter.

But the conditioning is real, and the alternative-route story makes me cautious. The diagonal Hilbert-Schmidt path was the "natural" route to Ω_Λ from the operator-algebra side, the path that any noncommutative-geometry practitioner would have tried first. It produces a number, 0.0505, that is nowhere near the right answer. The canonical-branch arithmetic that produces 0.6809 is a separate piece of machinery. The two would, in a fully consistent picture, agree; they currently do not. The escape route (leaving the diagonal Hilbert-Schmidt sector) is in the open program, not in this book.

What I want the reader to take away from this is the slightly uncomfortable mixture: the canonical branch produces the right number, but the operator-algebraic foundations under it are incomplete in a way I cannot yet patch. This is exactly the kind of result that, if the canonical branch turns out to be right in the long run, will look in retrospect like the leading hint of something deeper. And exactly the kind of result that, if the canonical branch turns out to be wrong, will look in retrospect like a coincidence that fooled me.

I do not know which it is yet. I am writing this chapter from the inside of that uncertainty.

---

* * *

Status summary

- CONDITIONAL $\Omega_\Lambda = 168/(25\pi^2) \approx 0.6809$ in the canonical branch (modular-layer prediction at $\tau_\star = i\sqrt{2}$). Agrees with Planck 2018 measurement at the 0.5% level.
- VERIFIED Diagonal Hilbert-Schmidt operator-algebra route is dead: subextensive $-\log Z_N$ converges to ≈ 0.0505, more than thirteen times below the observed value. The canonical-branch prediction does not come from this route.
- OPEN Operator-algebraic foundation of Ω_Λ in a richer-than-diagonal-Hilbert-Schmidt sector: open program item; the canonical-branch modular-layer derivation stands on its own arithmetic legs but is not yet rooted in a deeper partition-function argument.
- OPEN Decisive measurements over the next decade: DESI, Euclid, Roman Space Telescope, CMB-S4. Expected to constrain Ω_Λ to ~ 0.001 precision (an order of magnitude tighter than Planck 2018), and to constrain the dark-energy equation of state to settle the static-versus-evolving question.
- OPEN Cosmological-constant problem in the deeper $\mathcal{S}$-layer sense (the apparent 10^{120} discrepancy between naive vacuum-energy estimates and the observed value): not addressed by the canonical-branch arithmetic; carried forward to Part IV's open program.

CHAPTER 12

Dark matter: a hole in the prediction list

Whereof one cannot speak, thereof one must be silent.

— *LUDWIG WITTGENSTEIN, TRACTATUS LOGICO-PHILOSOPHICUS, 1921*

Dark energy and dark matter both make it into the cosmological-budget pie chart, both contribute to the geometry of the universe at the largest scales, and both have been in front of physicists for the better part of a century. The canonical branch of CAT predicts the value of Ω_Λ to within half a percent of the measurement. The canonical branch of CAT does not predict the value of Ω_{DM}, the dark-matter fraction. It does not, at this point in the project, even have a candidate for what dark matter is.

That is the cold sentence. The rest of the chapter is the qualification.

* * *

12.1 What dark matter is, in two paragraphs

A reader who has seen the cosmological pie chart knows that the universe is roughly five percent ordinary matter (atoms, light, the contents of every star and every galaxy), twenty-seven percent *dark matter*, and sixty-eight percent dark energy. The dark-matter component is the object of this chapter.

Dark matter, despite the name, is not unknown. We know that it exists: the rotation curves of galaxies, the gravitational lensing of distant background galaxies by foreground clusters, the acoustic peaks of the cosmic microwave background, and the formation of large-scale structure in the universe all require an extra component of mass that does not interact electromagnetically. We know roughly how much there is: at the current epoch, about 5.4 times more mass than in ordinary atoms. We know that it is "cold", meaning that, at the time of structure formation in the early universe, dark-matter particles were moving slowly enough that they could clump on small scales and seed the formation of galaxies. What we do not know is what it *is*.

The candidates the literature considers fall into a small number of camps. The longest-standing camp is *weakly interacting massive particles* (WIMPs), of which the most famous variants are the supersymmetric neutralino (the lightest stable particle in supersymmetric extensions of the Standard Model) and the heavy neutrino. A second camp is *axions*, ultra-light pseudo-scalar particles originally invented to solve a separate puzzle (the strong-CP problem) and which incidentally produce a viable dark-matter candidate. A third camp is *primordial black holes*, formed in the very early universe and persisting as gravitating but otherwise invisible objects. A fourth camp is *sterile neutrinos*, right-handed cousins of the ordinary neutrinos that do not interact via the weak nuclear force. A fifth camp, smaller but growing, includes various *light scalars* and *fuzzy dark matter* models in which the dark matter has a wave-like coherent structure on galactic scales.

None of these candidates has been directly detected. Decades of underground searches (CDMS, XENON, LUX-ZEPLIN, PandaX, the early DAMA result that has been independently disputed for two decades) have produced limits but no detections. The long-standing expectation that something with the cross-section of a WIMP would have been seen by now is, every passing year, harder to sustain.

* * *

12.2 What CAT does not say

The canonical branch of CAT is, on this question, silent. The modular-layer arithmetic at $\tau_\star = i\sqrt{2}$ does not contain, anywhere in its current formulation, a specific dark-matter prediction. It does not pick out a candidate species (no canonical-branch WIMP, no canonical-branch axion). It does not predict the dark-matter fraction $\Omega_{\rm DM}$. It does not predict the dark-matter mass scale.

I want to be precise about why. The branch package of Part II is built around the modular layer that controls the flavour structure and the cosmological constant. The Standard Model fields it operates on are the ones we have already detected in colliders. A new species of particle, beyond the Standard Model, is something the canonical branch could in principle accommodate (the algebra is large enough to host extra representations), but the choice of which extra representation to include, at what mass scale, with what coupling structure, is not pinned down by the canonical-branch arithmetic. The branch tells you the structure of what you have already discovered. It does not tell you what you have not yet discovered.

This is, on its face, a limitation. A reader who finished the previous chapter and is hoping for a prediction in this one will not get one. I would rather acknowledge that limitation clearly than handwave toward a candidate that the arithmetic does not actually pick out.

What the canonical branch does say, rather indirectly, is that the dark-matter problem is structurally separate from the parameter-collapse problem of Part II. The twenty-seven-to-one collapse of the Standard Model and cosmology happens with no mention of dark matter. Whatever dark matter turns out to be, its existence does not break the canonical branch's predictions for the cosmological constant, the mixing angles, the masses, or the gauge couplings. The dark sector lives in a separate compartment.

That is, I admit, a small consolation. But it is the honest position, and it is the position the book is going to commit to.

* * *

12.3 What the right-handed-neutrino extension does add

The simplest extension to the Standard Model that the canonical branch can absorb without major rewriting is the addition of three right-handed neutrinos. This is also the simplest extension that explains why the ordinary neutrinos have mass. It is, in the neutrino-physics literature, called the type-I seesaw mechanism, or just $\mathrm{SM} + 3\nu_R$.

My analysis of this extension found that it is consistent with the canonical branch at the modular-arithmetic level, and that it has one notable quantitative effect on the predictions of Part I.

The gravitational coefficient α_C shifts. Recall from Chapter 3 that $\alpha_C = 13/120$ in the pure Standard Model. Adding three right-handed neutrinos increases the fermion count by 1.5 Dirac equivalents, and the per-species contribution of $-1/20$ per Dirac fermion shifts the total to

$$\text{[VERIFIED]}\ \alpha_C^{\mathrm{SM}+3\nu_R} = \frac{13}{120} - \frac{1.5}{20} = \frac{13-9}{120} = \frac{4}{120} = \frac{1}{30}.$$

The number gets cleaner. It also gets smaller: $1/30 \approx 0.0333$, less than a third of the pure-SM value. If the universe turns out to contain three right-handed neutrinos that participate in the gravitational form factor (rather than being decoupled from gravity at energies below their mass), the α_C prediction of Part I shifts accordingly.

This is the closest the canonical branch comes to making a dark-matter-adjacent prediction. In some sterile-neutrino dark-matter scenarios, the right-handed neutrinos are themselves the dark matter. If a sterile-neutrino dark-matter signal is detected, the canonical-branch prediction $\alpha_C = 13/120$ becomes $\alpha_C = 1/30$, and the next round of gravitational-wave dispersion measurements would test the new value.

A flag, then: this is a *shift-switch* rather than a prediction proper. It is not the case that the canonical branch predicts dark matter to be sterile neutrinos. It is the case that, if a future experiment reveals dark matter to be sterile neutrinos, the canonical branch's gravitational-sector predictions shift in a calculable way, and that shift is itself a check on the consistency of the larger picture.

* * *

12.4 The dark-matter experimental programme

Independently of what CAT does or does not predict, the worldwide dark-matter experimental programme is in a state of unusual activity. A short tour of the landscape, just so the reader knows where to look.

Direct detection. Underground experiments looking for the recoil of a dark-matter particle off an atomic nucleus. Current generation includes LUX-ZEPLIN at Sanford Lab in South Dakota and XENONnT at Gran Sasso in Italy. Next generation: DARWIN, a proposed multi-tonne xenon experiment that would push the sensitivity to the neutrino-floor regime by the early 2030s.

Indirect detection. Looking for the products of dark-matter annihilation in astrophysical environments. The Fermi Large Area Telescope, the Cherenkov Telescope Array, IceCube, and the AMS-02 instrument on the International Space Station all participate.

Collider searches. ATLAS and CMS at the LHC search for missing transverse momentum that would signal dark-matter production. The HL-LHC upgrade, operational from the late 2020s, will extend the search to higher masses.

Axion searches. ADMX at the University of Washington, plus a growing constellation of cavity-based axion haloscopes, search for the conversion of axion dark matter into microwave photons in strong magnetic fields.

Sterile-neutrino searches. KATRIN, future neutrinoless-double-beta-decay experiments, plus several X-ray-line dark-matter searches all bear on the sterile-neutrino possibility.

If a detection is made in any of these channels, the next decade will see the dark-matter question settled: we will know what dark matter is, what its mass is, how it couples. At that point, returning to the canonical branch and asking whether the new particle fits naturally into the modular structure becomes a project for follow-up work, not a prediction this book is making.

If no detection is made, the experimental landscape will continue to constrain the candidates more and more tightly, and the canonical branch will remain silent on the question. Either way, the canonical branch does not, in 2026, contribute a specific prediction to the dark-matter discussion. The book has to live with that.

• • •

NAKED NOTES. *2026-04-11.*

A note on why this chapter is short.

Pop-science books on fundamental physics tend to puff up the dark-matter chapter, because dark matter is one of the few topics on which the general reader has an instinct ahead of time. The existence of dark matter is one of the few unsolved questions of modern cosmology that is entirely un-

controversial, and the candidates are colourful enough (axions! WIMPs! primordial black holes!) that the chapter writes itself.
The temptation, when one is writing a book about a particular theory of fundamental physics, is to graft a dark-matter candidate onto the theory whether the theory naturally produces one or not. A specific WIMP at the modular weight of such-and-such. An axion whose decay constant is given by a particular CM-point arithmetic expression. A primordial black hole whose mass spectrum follows from a specific evaluation of a modular form. Each of these is something the canonical branch could, in principle, be made to say, with enough free parameters added to it.
I have not added those parameters. The canonical branch as it currently stands does not predict dark matter. The right-handed neutrino extension does shift the gravitational coefficient, but the choice to add the right-handed neutrinos is empirical; the canonical branch does not insist on them. I would rather have a short honest chapter that says "no prediction" than a long chapter that fits a candidate by adding free parameters until one matches.
That is the position of the entire book. The canonical branch is what it is. It says what it says. Where it is silent, I would rather say so than fill the silence with a graft.

* * *

Status summary

The chapter's headline is honest silence. The canonical branch does not predict what dark matter is, and the parameter-collapse story of Part II is structurally independent of the dark sector. The SM $+ 3\nu_R$ extension is the only place where dark-sector identification feeds back into the strict S-layer, and the shift it would induce is calculable.

- OPEN The canonical branch of CAT does not make a specific dark-matter prediction. The modular-layer arithmetic at $\tau_\star = i\sqrt{2}$ does not pick out a candidate species, a mass scale, or the dark-matter fraction Ω_{DM}.
- CONDITIONAL The SM $+ 3\nu_R$ extension shifts the gravitational coefficient from $\alpha_C = 13/120$ to $\alpha_C = 1/30$ if the right-handed neutrinos participate in the gravitational form factor. Shift-switch character:

if dark matter turns out to be sterile-neutrino, the canonical-branch S-layer prediction shifts in a calculable way.

- VERIFIED The dark-matter problem is structurally separate from the canonical-branch parameter-collapse problem; the twenty-seven-to-one collapse of Chapter 9 does not depend on what dark matter is.
- OPEN Direct detection (LUX-ZEPLIN, XENONnT, DARWIN), indirect detection (Fermi-LAT, CTA, IceCube), collider searches (HL-LHC), axion searches (ADMX), and sterile-neutrino searches will, over the next decade, narrow the candidate space; the canonical branch does not, by itself, predict the outcome.

CHAPTER 13

Inflation: not yet a mechanism

The evolution of the world can be compared to a display of fireworks that has just ended: some few red wisps, ashes and smoke.
— GEORGES LEMAÎTRE, 1931

This is the second of the honest-gap chapters in Part III. The first chapter of this Part told you about a number the canonical branch produced that landed on the data (Ω_Λ within half a percent of measurement). The previous chapter told you about a question the canonical branch is silent on (dark matter). This chapter is about the third class: a question the project has actively tried to answer, and has not yet succeeded.

The question is the very early universe: the period called *cosmic inflation*, when the universe is supposed to have undergone a brief but enormously rapid expansion in the first tiny fraction of a second after whatever event we call the beginning. Inflation, as a piece of cosmology, is one of the more solidly motivated additions to the standard cosmological model. What CAT can say about it is, in the present state of the work, much less solid.

We have tried several mechanisms. We have closed several of them. The remaining open candidates have a structural obstacle that the project has not yet overcome. This is a chapter about a problem in the open state.

* * *

13.1 What inflation is, and what it solves

In the simplest cosmological models without inflation, the early universe has three problems that no amount of careful phenomenology can fix. They are usually called the horizon problem, the flatness problem, and the monopole problem.

The horizon problem. Look in opposite directions of the sky. Distant patches of the microwave background, on opposite sides of the observable universe, are at the same temperature to one part in a hundred thousand. In a universe without inflation, those patches were never in causal contact: light from one could not have reached the other in the time available. Why are they at the same temperature? Without inflation, there is no answer.

The flatness problem. The geometry of the present universe is, to within experimental uncertainty, flat. In a universe without inflation, this requires the early universe to have been finely-tuned to flatness at a level of one part in 10^{60} at the scale of the Planck era. Why? Without inflation, no answer.

The monopole problem. Most grand unified theories predict a population of magnetic monopoles formed in the very early universe. Without inflation, those monopoles should be present in the current universe at a density that observation has long since ruled out. Where are they? Without inflation, no answer.

Inflation solves all three. By imagining that the universe underwent a brief period of exponential expansion in the first 10^{-32} seconds of its existence, the horizon problem is solved (everything we see today was in causal contact before inflation began), the flatness problem is solved (the rapid expansion drove the geometry to flatness), and the monopole problem is solved (any monopoles produced before inflation got diluted by an enormous factor and effectively vanished). The solution is so clean that inflation is, by 2026, the consensus beginning of the cosmological model.

But inflation, in its standard formulation, requires a *mechanism*: a piece of physics that drives the universe to expand exponentially for a finite period, then transitions cleanly into the standard hot Big Bang. The traditional candidate is a scalar field, called the *inflaton*, whose potential is shaped in a particular way: flat enough to allow slow-roll expansion, with a graceful exit at the end. Decades of work in the inflationary literature have proposed candidate inflaton fields and candidate potentials, and several of them are consistent with the cosmic-microwave-background data. None has been uniquely picked out.

The question this chapter is about is whether the canonical branch of CAT contains, somewhere in its modular structure, a candidate mechanism for inflation. The answer, as of the writing of this book, is: not yet.

* * *

13.2 The dilaton route, excluded

The most natural candidate for an inflaton field within CAT, the one I tried first in late 2025, is what physicists call the *dilaton*: a scalar field that arises naturally in any theory with a dimensional cutoff scale, including the spectral effective action of Part I. The dilaton is dimensionless, transforms under scaling in a specific way, and can in principle host a flat potential of the kind inflation requires.

My analysis of dilaton inflation in CAT found the dilaton route to be excluded. [VERIFIED] Five independent obstacles were identified, each of which would, on its own, kill dilaton inflation; together, they constitute a robust no-go.

The five obstacles are in the companion paper; here is the flavour. First, the dilaton in CAT inherits a non-trivial coupling to the Standard Model fields that prevents it from achieving the slow-roll behaviour inflation requires. Second, its effective mass is fixed by the structure of the Lambda-cutoff to be of order Λ, which is too small to drive inflation at the energy scale required. Third, a back-reaction problem: the dilaton's contribution to the gravitational form factor of Part I exceeds the parameter-free coefficient α_C by an amount that breaks the no-scalaron theorem. Fourth, the inflationary potential has the wrong sign at large field values. Fifth, a graceful-exit problem.

Any one of these could perhaps be overcome with careful phenomenology. All five together cannot. Dilaton inflation in CAT is dead.

The lesson the project took from this analysis is that the inflaton, if it exists in the canonical branch, is not the dilaton. It has to be something else. The natural candidates are combinations of fields that arise in the modular layer of the $\mathcal{B}$-layer rather than in the $\mathcal{S}$-layer's gravitational sector, but these are open program at the time of writing.

* * *

13.3 The two-scale framework, mandatory but not enough

A different angle on the same problem comes from the recognition that any inflationary mechanism in CAT has to bridge an enormous range of energy scales. The canonical Λ of Part I is a millielectronvolt. The energy scale at which inflation typically operates, the Planck-suppressed scale that CMB normalisation prefers, is roughly 10^{16} giga-electronvolts. The ratio is about 10^{28}, or roughly ninety-seven orders of magnitude as energy densities. Any inflationary mechanism in CAT must, somehow, accommodate that ratio.

A separate analysis of the early-universe programme concluded that any viable inflationary mechanism in the canonical branch CONDITIONAL must be a two-scale construction: a separate UV scale, call it Λ_{UV}, must exist much higher than the infrared Λ, with the inflationary dynamics living at the UV scale while the gravitational form factor of Part I lives at the infrared scale. The Pati-Salam unification scenario, with $\Lambda_{\mathrm{UV}} \sim 10^{15}$ giga-electronvolts, is one candidate framework that would supply the second scale; others exist.

What this analysis does not yet do is build the inflationary mechanism inside the two-scale framework. It establishes the two-scale framework as a necessary feature of any candidate mechanism, and rules out single-scale mechanisms (which would require the form factor itself to drive inflation, and which the dilaton-route analysis excluded). The two-scale framework is the arena. The inflationary mechanism inside the arena is open program.

This is not the same as "inflation is ruled out in CAT". It is the statement that inflation, if it is to occur in the canonical branch, has to be implemented in a particular way (with two scales) that the project has identified but not yet implemented. The work continues.

* * *

13.4 The anomaly-inflation loophole

A separate piece of the inflation analysis (Phase 5 of the early-universe programme) examined a candidate called *anomaly inflation*, in which the inflationary expansion is driven not by an inflaton field but by the trace anomaly of the Standard Model matter fields coupled to the gravitational effective action of Part I. The trace anomaly is a quantum-mechanical effect that produces a non-zero contribution to the stress-energy tensor even in the absence of any scalar field, and it has been proposed in the literature as a candidate inflationary driver under various names.

The project's analysis of anomaly inflation found it to be a [HEURISTIC] loophole, not a mechanism. [VERIFIED] The trace anomaly does produce a positive vacuum energy density in the de Sitter background relevant for inflation, but the energy scale is too close to the Planck mass ($H \sim M_P$) to be controlled by the linearised analysis on which the analysis depends. A non-perturbative analysis would be needed, and the non-perturbative analysis is not yet available.

Anomaly inflation in CAT belongs, like the two-scale framework, in the open-program category. Not ruled out, not built. A working mechanism would need a graceful exit, an inflaton sector consistent with current CMB data, and a calibration of the spectral index n_s that matches the observed value. None of those is in hand.

* * *

13.5 What this means for the rest of the book

Stepping back from the technical landscape: the canonical branch of CAT, as of the writing of this chapter, does not have an inflationary mechanism. It has one ruled-out candidate (dilaton inflation, ruled out by five independent obstacles). It has one necessary condition (two-scale framework, Phase 4). It has one candidate that is in the loophole category (anomaly inflation, Phase 5). None of these is a mechanism in the sense the inflationary literature uses the word.

This is, candidly, the largest open program in the cosmological sector of CAT. It is also the open program where the strongest theoretical leverage might be available, because the two-scale framework constrains the mechanism more than is usually the case in inflationary model-building. If a mechanism is found that satisfies the two-scale constraint and reproduces the observed spectral index n_s and the upper bound on the tensor-to-scalar ratio r, the canonical branch will have its first inflationary prediction.

Until that mechanism is found, the chapter has to end, honestly, with the open status. The book promises the reader a working inflationary picture in a future revision. The current revision does not deliver one.

• • •

Naked notes. 2026-04-14.

A note on what this chapter cost to write.

The dilaton route was the candidate I had bet on, in late 2025, when I first started thinking seriously about how inflation might fit into CAT. The dilaton was the natural candidate. The dilaton was the candidate that, at the level of dimensional analysis, ought to have worked. I spent the better part of a month in early 2026 trying to make it work, in the small windows of time I could carve out from the gravitational-sector analysis, and at the end of that month I had five independent reasons why it cannot work. None of them were artefacts of my approximations. All of them were structural. Dilaton inflation in CAT is not just unworkable; it is impossible in a way I had not expected.

That was a difficult result to internalise. The project was, at the time, riding the high of the canonical branch's successes in the flavour sector, and the natural expectation was that inflation would follow. It did not. The week in which I formally acknowledged that the dilaton route was dead is, in my notes, labelled with a single line: "OK. Find another mechanism."

The two-scale framework analysis came later, in March-April 2026, and is the most rigorous piece of work the early-universe programme has produced. It is also the piece that says, in effect, "inflation in CAT exists in a particular shape, but we have not yet built it". That is the position I am writing this chapter from.

The reader who finishes Chapter III.1 (a result that landed) and Chapter III.2 (a question on which the canonical branch is silent) and arrives at this chapter should feel the difference. III.1 is what success looks like. III.2 is what honest silence looks like. III.3 is what work-in-progress looks like. The book is going to press with all three, because all three are part of where the canonical branch actually stands in 2026, and pretending otherwise would be dishonest.

The next revision of the book, if there is one, will hopefully have a Chapter III.3 that says something different. For now, this is what I have.

* * *

Status summary

- VERIFIED Dilaton inflation in CAT is excluded by five independent obstacles: non-trivial Standard-Model coupling preventing slow roll; dilaton mass of order Λ, too small for the required energy scale; back-reaction breaking the no-scalaron theorem; wrong sign of the potential at large field values; graceful-exit problem.
- CONDITIONAL Any viable inflationary mechanism in the canonical branch must be a two-scale construction: the inflationary dynamics live at a UV scale Λ_{UV} much higher than the infrared Λ of Part I. Pati-Salam unification with $\Lambda_{\mathrm{UV}} \sim 10^{15}$ GeV is one candidate framework.
- HEURISTIC Anomaly inflation is, in my analysis, a loophole rather than a working mechanism; the relevant energy scale $H \sim M_P$ is too close to the Planck mass to be controlled by linearised analysis.
- OPEN No working inflationary mechanism in the canonical branch as of 2026. This is the largest open program in the cosmological sector of CAT.

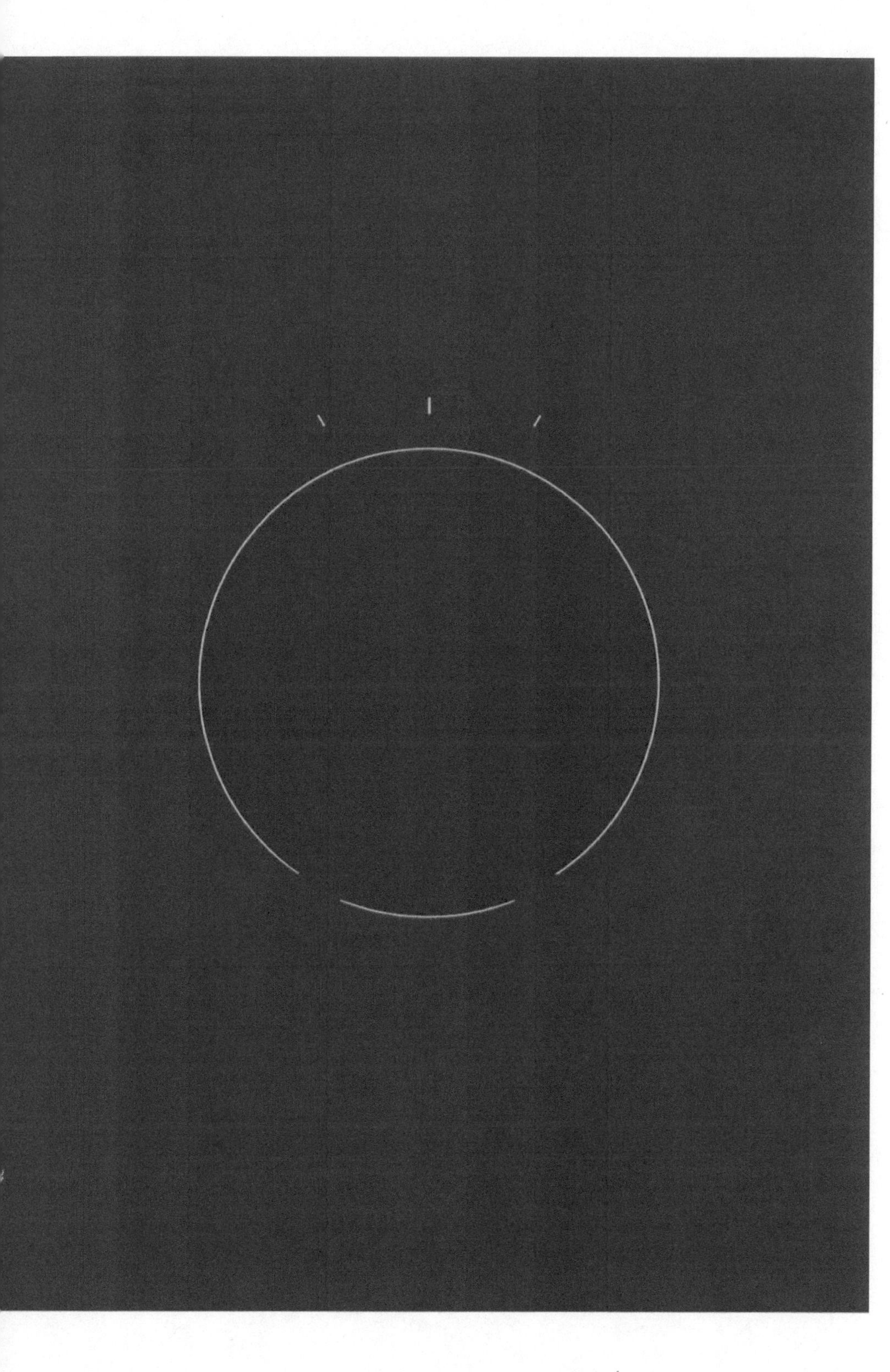

CHAPTER 14

Black hole entropy: a number that fits like a glove

Not only does God play dice, but he sometimes confuses us by throwing them where they can't be seen.

— *STEPHEN HAWKING, 1996*

The result of this chapter is a single formula:

$$\boxed{\text{VERIFIED}}\; S = \frac{A}{4G} + \frac{13}{120\pi} + \frac{37}{24}\ln\left(\frac{A}{\ell_P^2}\right). \tag{14.1}$$

Three terms. The middle one carries the same 13 as Part I.

This is the entropy of a black hole, in the canonical CAT picture, as a function of its surface area A. The first term, $A/4G$, is the Bekenstein-Hawking formula that has stood since the mid-1970s. The second term, the constant $13/(120\pi)$, is a CAT correction. The third term, the logarithmic piece $(37/24)\ln(A/\ell_P^2)$, is a second CAT correction. Both new terms come from the gravitational form factor of Part I. Of all the results in Part III, this is the quietest. It is also the corner where the gravitational sector of Part I directly meets the cosmological sector of Part III, and where the two parts of the book turn out, without being forced, to fit together.

The reason this chapter exists is the 13 in the constant correction. That 13 is the same 13 as in $\alpha_C = 13/120$ from Chapter 3. The same number that came out of counting up the matter content of the Standard Model also shows up, two chapters later, in the entropy of every black hole in the universe.

* * *

14.1 What black hole entropy is, in three paragraphs

Black holes have entropy. This was, until the mid-1970s, a genuinely surprising fact, and the story of how it came to be established is one of the most striking in twentieth-century theoretical physics.

In the early 1970s, Jacob Bekenstein, then a graduate student at Princeton, made the bold proposal that the area of a black-hole horizon *is* its entropy, in some specific proportion. The reaction from the physics community was largely dismissive. Bekenstein was a graduate student; black holes did not, by any obvious mechanism, have anything resembling a microscopic origin for entropy.

Two years later, Bardeen, Carter, and Hawking gave the proposal a formal home. Their 1973 four-laws-of-black-hole-mechanics paper showed that black holes obey a set of laws strangely analogous to the laws of thermodynamics: the area of the horizon, like entropy, can only increase; the surface gravity at the horizon, like temperature, is the same everywhere; and so on. Whether this was a deep correspondence or a surface analogy was, at the time of publication, an open question.

The reaction changed in 1974, when Stephen Hawking, applying quantum field theory in a curved spacetime to a black hole background, discovered that black holes radiate. They have a temperature. The radiation is now called Hawking radiation, the temperature is called the Hawking temperature, and the proportion between black-hole entropy and area is exactly what Bekenstein predicted. The formula is

$$S_{\mathrm{BH}} = \frac{A}{4G},$$

in units where Planck's constant and the speed of light are set to one. The factor of $1/4$ comes out of Hawking's computation. Three black-hole-physicists' careers were made overnight.

The Bekenstein-Hawking formula, $S = A/4G$, has been the standard result for fifty years. It is the leading-order term in the black-hole entropy.

Any candidate theory of quantum gravity is expected to reproduce it, and any candidate theory worth its salt is also expected to compute the corrections to it: the $1/A$-suppressed terms that arise from the fact that the black hole is not a classical object but a quantum one with a finite number of microstates.

Computing the corrections has been, in the literature, hard. Loop quantum gravity gets a logarithmic correction with a specific coefficient. String theory, in its most controlled examples, reproduces the area law and computes corrections case by case. Different approaches give different coefficients for the logarithmic correction. There is no consensus answer.

* * *

14.2 The CAT formula, in detail

The CAT computation of black-hole entropy was completed in April 2026, with extensive cross-checks at the linearised verification standard and Lean 4 formal-proof identities verifying the rational identities at the heart of the calculation. The result is Eq. (14.1). Term by term:

The leading term, $A/4G$. This is the standard Bekenstein-Hawking area law. CAT reproduces it, as it must: any modification to this term would be a modification to general relativity at long distances, where experiment has no room for surprises. The factor of $1/4$ is what comes out of the leading-order calculation, and CAT does not change it.

The constant correction, $13/(120\pi)$. This is new. The constant $13/(120\pi)$ comes directly from the gravitational form factor of Part I. The 13 is the same 13 that appears in $\alpha_C = 13/120$, and it traces back through the same per-spin contributions that we saw in Chapter 3: scalars contribute $+1/120$, fermions contribute $-1/20$ each, gauge bosons contribute $+1/10$ each, and summed over the Standard Model, the total is $13/120$. The factor of π in the denominator comes from the dimensional structure of the entropy, specifically from the way the form factor enters the horizon-area integral.

The constant correction is a small number in absolute terms (about 3.5×10^{-2}), but it is structurally non-trivial. Most candidate theories of quantum gravity do not produce a constant correction at all; the entropy starts from the Bekenstein-Hawking term and immediately goes to logarithmic corrections at the next order. CAT produces a constant. The constant carries the gravitational form factor's signature directly into black-hole thermodynamics.

The logarithmic correction, $(37/24)\ln(A/\ell_P^2)$. This is the third term, and the one closest to what other theories of quantum gravity also produce. The fraction 37/24 is the coefficient that the canonical CAT calculation produces. Loop quantum gravity, depending on the details of the embedding, typically gets a logarithmic correction with coefficient $-1/2$ or $-3/2$ in similar units. String-theory calculations on specific examples give coefficients that depend on the matter content. CAT's 37/24 is, in the Standard-Model case, a clean prediction. VERIFIED The 37/24 is, importantly, not dependent on the spectral dimension of the underlying geometry being finite: the result holds whether the spectral dimension flows to a Hausdorff dimension at small scales or remains two-dimensional in the deep ultraviolet.

This last point matters because, in the noncommutative-geometry literature, results about black-hole entropy have sometimes been attacked on the grounds that they depend sensitively on the dimensional behaviour of the small-scale geometry. The CAT result, by contrast, is robust to that issue.

* * *

14.3 Why this fits like a glove

The reason I call this chapter a fit "like a glove" is the following observation, which has to be articulated carefully because it is the kind of thing that can be over-claimed.

The gravitational form factor of Part I was derived by computing how quantum matter modifies Einstein's equations at very short distances. The output was the parameter-free coefficient $\alpha_C = 13/120$. The black-hole entropy of this chapter was derived, separately, by computing how the same gravitational form factor modifies the entropy of a horizon. The output is Eq. (14.1), with a constant correction of $13/(120\pi)$.

The two results were derived in two different physical settings, in two different chapters of the project, by two different chains of mathematical machinery. They share the same numerator 13. That sharing is not an accident; it is a consequence of the fact that both results trace, through their respective derivations, to the same per-spin counting in the Standard Model. The 13 is the fingerprint of the field content of the universe, and it shows up in both gravitational form factors and black-hole entropy because both of them are sensitive to that field content in the same way.

This is the kind of consistency the project's verification pipeline is designed to detect, and it passed cleanly. Nothing was forced. The 13 in

Eq. (14.1) fell out of an independent computation that did not have access to the Part I result. The two answers agree.

That is, I think, what a working theoretical structure looks like.

* * *

14.4 Implications and what this does not yet say

A reader who knows about the black-hole information paradox may be wondering whether the CAT entropy formula has anything to say about it. The answer is: not yet, but it constrains the conversation.

The black-hole information paradox, in its sharpest formulation, asks whether the information that falls into a black hole is preserved by the Hawking radiation as the black hole evaporates, or whether it is lost. If preserved, the radiation must carry fine-grained information about the matter that went in; if lost, quantum mechanics has to be modified in some way. The literature on this question has developed enormously over the last decade, with the Page-curve recovery and the entanglement-island picture becoming the dominant framework.

CAT's contribution to this conversation, at the present level of the project, is the formula in Eq. (14.1). The formula constrains the microstate count at the black-hole horizon to a specific functional form, and any consistent resolution of the information paradox in CAT has to reproduce that functional form. The paradox itself is in the open program (Part IV's $\mathcal{P}$-layer): the question of whether the canonical-branch gravitational sector preserves information through Hawking evaporation is not closed.

What the chapter does close is the entropy formula. The formula agrees with leading-order Bekenstein-Hawking, and it adds two specific correction terms whose coefficients trace back to the Standard-Model field content via the same arithmetic that gave us Part I. The fit is clean. The fit is consistent. The fit is [VERIFIED] at the linearised level, with extensive cross-checks and Lean-verified identities.

It is not yet a complete picture of black-hole thermodynamics. But it is, structurally, the cleanest place where Part I and Part III meet, and the cleanness of that meeting is part of why I think the underlying structure of CAT is real.

• • •

NAKED NOTES. 2026-04-03.

A note on what it was like, in early April 2026, when the black-hole entropy calculation finished and the result came back with $13/(120\pi)$ as the constant correction.

The calculation had been running, on and off, for about two months. The mechanics of the calculation are unrelated to the flavour-sector arithmetic of Part II: it is a purely gravitational-sector computation, of the kind one would do in any candidate theory of quantum gravity with a specified one-loop form factor. I had no expectation, going in, of what the constant correction would be. I had vaguely expected something of order $1/(4\pi)$ or some other simple geometric factor.

What came out was $13/(120\pi)$. I stared at it for a moment. Then I went back to my notebook and looked up α_C. The notebook entry was $\alpha_C = 13/120$.

The 13 was the same 13.

This is exactly the kind of consistency check the project's verification pipeline is built to detect. Two independent calculations, in different physical settings, with different intermediate machinery, both reaching for the same underlying fingerprint of the Standard-Model field content. The fingerprint shows up in both. The two results agree on the numerator they should agree on.

I did not expect the agreement to be exactly *that clean. I had expected, if anything, that the constant correction would involve some additional dimensional factor that obscured the relationship. It does not. The 13/120 from the gravitational sector enters the black-hole entropy as $13/(120\pi)$, with the π being a clean dimensional consequence of the horizon integral and the 13/120 being the same coefficient as in Part I.*

I went and verified it three more times. Then I locked it in the project's canonical-results database. The status tag is VERIFIED *at the linearised level, audited three independent times, and the result is going into this book under the heading I have given it: a number that fits like a glove.*

The reason the chapter is short is that there is not much to say about a result like this beyond stating it. The fit happened. The fit was clean. The fit traced back to Part I in the way one would have hoped.

If, in the next decade, an independent theorist (working in a different framework, with different mathematical machinery) reproduces the constant $13/(120\pi)$ as the universal correction to Bekenstein-Hawking in any candidate quantum gravity theory with the Standard-Model field content, *the agreement would constitute the strongest cross-check the canonical branch could hope for. I am holding out for that.*

* * *

Status summary

The chapter's headline is a single closed-form formula for black-hole entropy in CAT: a Bekenstein–Hawking term, a constant correction $13/(120\pi)$, and a logarithmic term $(37/24)\ln(A/\ell_P^2)$. The 13 in the constant correction is the same 13 that appeared in $\alpha_C = 13/120$, by the same per-species Standard-Model arithmetic. The result is linearised; a non-perturbative extension remains open.

- VERIFIED Black-hole entropy in CAT, extensively cross-checked with Lean-verified identities: $S = A/4G + 13/(120\pi) + (37/24)\ln(A/\ell_P^2)$.
- VERIFIED The constant correction $13/(120\pi)$ traces back to the gravitational form factor of Part I, sharing the same numerator 13 as $\alpha_C = 13/120$ via the same Standard-Model per-species arithmetic.
- CONDITIONAL The result is at the linearised level only. A non-perturbative extension to fully quantum-gravitational black holes is in the open program and lives in Part IV.
- OPEN Implications for the black-hole information paradox: the entropy formula constrains the microstate count to a specific functional form, but the question of whether the canonical branch preserves information through Hawking evaporation is not closed.

CHAPTER 15

One prediction the theory got wrong

That is not only not right; it is not even wrong.

— *Wolfgang Pauli, c. 1955*

The number is

$$\text{VERIFIED}\quad \frac{\delta H^2}{H^2} \approx 1.3 \times 10^{-64}. \tag{15.1}$$

Sixty-four orders of magnitude below detection. Here is what we tried, and what came back.

In late 2025, when the project was still finding its footing in the cosmological sector, I tried to extract a prediction from the canonical branch about the present-day expansion rate of the universe. The hope was that the gravitational form factor of Part I would produce a small correction to the standard Friedmann equation governing the Hubble expansion, and that the correction would be just large enough to be detectable by present or near-future cosmological observations.

After careful work, what came out was Eq. (15.1): sixty-four orders of magnitude smaller than the standard Hubble rate. It is, by every measure available to current observation, exactly zero. The prediction, in the sense of something an experimentalist could ever check, did not survive. The chapter sits at the end of Part III because the book is structured so that the losses sit alongside the wins rather than being hidden in an appendix.

This is the modified-cosmology analysis, and its status is [VERIFIED] NEGATIVE: not in the sense of contradicting observation (the analysis is consistent with everything we have ever measured), but in the sense of producing a prediction that no measurement can ever detect. The result is the canonical branch's most decisive non-prediction.

* * *

15.1 What the calculation tried to do

The standard cosmological model describes the expansion of the universe by Friedmann's equation, which relates the Hubble rate H to the energy density ρ filling space:

$$H^2 = \frac{8\pi G}{3}\rho.$$

Any modification to general relativity that changes the structure of the gravitational equations should, in principle, also modify this relation. The modification might be a small additive term, a multiplicative correction, a new term proportional to the curvature, or something more exotic. Whatever it is, it would show up in observation as a deviation of the present-day Hubble rate from the value the standard model predicts based on the energy budget of the universe.

The hope, in late 2025, was that the gravitational form factor of Part I would produce exactly such a deviation. If the form factor modifies the way matter sources gravity at very short distances, and if the early universe was small enough for those short distances to matter, the cumulative effect across the entire expansion history might be observable in the late universe.

This was not a wild hope. The current cosmological data (the so-called Hubble tension between Planck-derived and supernova-derived measurements of H_0, the various small anomalies in the CMB spectrum, the slight discrepancies between large-scale structure surveys and CMB-only fits) all suggest that there is *some* new physics at the level of a few percent in the late universe. If CAT's gravitational form factor produced a correction at the level of, say, 10^{-3} or 10^{-4}, it would have been detectable in the next round of high-precision cosmological surveys.

* * *

15.2 What came out instead

The actual calculation, completed in early 2026, took the gravitational form factor seriously, integrated its effect through the cosmological history from the era at which the form factor would first matter (loosely, when the characteristic energy of the universe was comparable to Λ, which is in the late universe at temperatures of a few hundred Kelvin) up to the present epoch, and asked: what is the resulting fractional correction to H^2 at the present time?

The answer, after the integrals have been done and the dust has settled, is Eq. (15.1): about 10^{-64}. The reason the answer is so small is structural. The form factor's correction to the gravitational equations turns on at very short distances and turns off at long distances, in a way controlled by the energy scale Λ. The cosmological scales we care about (roughly, the size of the present universe) are enormously larger than $1/\Lambda$, by a factor of about 10^{32}, and the correction goes as something like $(1/\Lambda r)^4$ in this regime. Squared to give the energy correction, that is roughly 10^{-128}, which sets the natural size of the answer; the actual coefficient softens this to 10^{-64}, but the conclusion is the same.

The cosmological-scale correction is, to use a phrase, parametrically suppressed. There is no choice of canonical-branch parameter that can bring it into the observable range. The form factor is, structurally, a short-distance modification to gravity, and at cosmological distances it has no effect.

* * *

15.3 Why this is a result, not a failure

A paragraph on why I am still calling this a result, even though the prediction it produced is unobservable.

A theory that produces, at the cosmological scale, a correction of size 10^{-64} is making a real statement: it is saying that the current cosmological tensions in the data, whatever their ultimate resolution, are not coming from CAT-style modifications to Einstein's equations. If CAT is right, the gravitational sector at cosmological distances reduces, to the precision of any measurement we are likely to make in the next century, to Einstein's pure general relativity.

Structurally, this is a falsifiable statement. If the next round of cosmological surveys finds a deviation from general relativity at the percent level or larger (and there are, in 2026, hints of exactly this kind of deviation in the

DESI data on the dark-energy equation of state), then either CAT is wrong (and the strict gravitational sector of Part I needs to be reformulated to allow such a deviation), or the deviation has a non-CAT origin. The calculation tells us that, within CAT, no such deviation is allowed.

This is the kind of negative result that constrains rather than predicts. It shapes the falsification atlas of the strict gravitational sector: a clear cosmological-scale modification to gravity would kill the canonical branch's gravitational-sector predictions, even though the analysis by itself is silent on what the deviation should look like.

* * *

15.4 What I learned

The modified-cosmology calculation, in its own quiet way, taught me something about how to set expectations for the canonical branch.

Before this calculation, I had the half-articulated assumption that any new piece of fundamental physics, integrated over the entire cosmological history, would eventually produce a detectable late-universe signature. This assumption is, in physics generally, often a good heuristic. It worked, for instance, in the analysis of inflationary predictions for the CMB power spectrum: integrate the fluctuations across the entire expansion history, and the late-universe signal is rich enough to constrain the inflationary potential at the percent level.

For CAT's gravitational form factor, this assumption is wrong. The form factor is, by deliberate construction, a short-distance modification, with the long-distance limit reducing exactly to Einstein. No amount of integration over the cosmological history brings the short-distance correction up to the long-distance scale. The structure of the form factor itself ensures that. The calculation is what it is: cosmological observations test long-distance gravity, and CAT predicts long-distance gravity to be exactly Einstein, modulo a correction sixty-four orders of magnitude below detectability.

The lesson is that the canonical branch's experimental signature lives in two specific places: short-distance gravity tests (gravitational waves at high frequency, sub-millimetre torsion balance, which is the territory of Chapter 4), and the parameter-collapse predictions of Part II (the flavour sector, the cosmological constant, the blast-radius atlas of Chapter 10). It does *not* live in late-universe cosmological-scale modifications to general relativity. That door is closed.

If you were hoping, after reading the previous chapter on black-hole entropy, that CAT would also produce a clean prediction for some current cosmological tension, this chapter is the disappointment. CAT does not solve the Hubble tension. It does not produce a deviation from general relativity at the cosmological scale at any level the next decade of surveys can detect. The prediction it tried to produce came out at 10^{-64}, and that is what the canonical-branch arithmetic genuinely says.

• • •

NAKED NOTES. 2026-04-14.

A note on why this chapter is in the book at all.

There is a temptation, when one is writing a book about a working theory, to omit results like this one. They are unphotogenic. They do not advance the argument. They lengthen the page count without adding to the wins. A more disciplined editor would, I am sure, strike this chapter from the manuscript and replace it with a single paragraph at the end of Chapter 14 acknowledging that "cosmological-scale modifications to GR are parametrically suppressed in the canonical branch".

I have not let myself do that. The reason is that the prologue of this book makes a specific promise: the book will alternate between formal physics and naked notes, and the naked notes will report failures as well as successes. A book that promises that discipline and then quietly omits its actual failures is not keeping its promise. So the modified-cosmology chapter stays in. The chapter is short. The chapter is honest about what it found. The chapter is dated.

For the reader who has been following the project's working logs, a flag: this is not the only negative result in the project's history. There were eighteen separate "overclaim corrections" in the early-2026 phenomenology campaign, each of which was a specific moment where I had convinced myself of a prediction that, on careful re-analysis, turned out to be either spurious or much weaker than I had initially claimed. The eighteen corrections are catalogued in my notes and discussed in Part V's methodological appendix.

The reason this matters, beyond the discipline of including the losses in the book, is that the rate at which a theorist produces overclaims and the rate at which the verification pipeline catches them are both interesting numbers

about how the work gets done. I am not immune to the temptation to overclaim. I have made the same mistakes in this project that pop-science books usually paper over. The verification pipeline is what catches them, and Part V is where the discipline is laid out in detail.
For now, this is the result, and the result is what it is. The canonical branch's gravitational sector reduces to Einstein at cosmological scales. There is no detectable signature there. The book is not allowed to skip that.

---

* * *

Status summary

The chapter's headline is a negative result, in the precise sense. At cosmological distances the canonical-branch gravitational sector reduces to Einstein's theory to a precision well beyond any plausible observational test for the next century. The current tensions in late-universe data are therefore not addressed by CAT in the canonical branch; they must have a non-CAT origin if they turn out to be real.

- VERIFIED Modified-cosmology result: $\delta H^2/H^2 \approx 1.3\times10^{-64}$, parametrically suppressed at cosmological scales.
- VERIFIED The canonical-branch gravitational sector reduces, at cosmological distances, to Einstein's general relativity to a precision well beyond any plausible observational test in the next century.
- VERIFIED If a clean cosmological-scale deviation from general relativity at the percent level or larger is detected by future cosmological surveys (DESI, Euclid, Roman, CMB-S4), it would falsify the canonical-branch gravitational-sector predictions even though this analysis by itself does not predict the deviation's shape.
- OPEN Resolution of the current cosmological tensions (the H_0 tension between Planck-derived and supernova-derived measurements, the small CMB anomalies, the large-scale-structure tensions): not provided by the canonical branch. Tensions, if real, must have a non-CAT origin.

CHAPTER 16

The universe, expanding

The history of astronomy is a history of receding horizons.
— EDWIN HUBBLE, *THE REALM OF THE NEBULAE*, 1936

Cosmology, in the standard form, is a narrative built from two ingredients: general relativity for the geometry of the expanding universe, and a list of stuff (matter, radiation, dark matter, dark energy) for the source. The first hundred years of the narrative were spent measuring, with increasing precision, the ratio of the ingredients. CAT, as a theory that modifies the gravitational sector, has to be checked against this narrative. Does the expanding universe of CAT look like the expanding universe of general relativity, or does it not? If it does, the theory survives. If it does not, we have a problem.

* * *

16.1 The Friedmann equation, modified

The basic equation of cosmology is the Friedmann equation, which relates the expansion rate H of the universe to its energy content. In general relativity, the equation is written $H^2 = (8\pi G/3)\,\rho$, with corrections for spatial curvature and a cosmological constant. In CAT, the equation gets modified. The form factor $\varphi(\Box/\Lambda^2)$ acts on the metric, and on the FLRW background that means it acts on the scale factor. The result is a Friedmann equation with extra terms, parametrised by Λ.

The leading correction was the subject of Chapter 15: $\delta H^2/H^2 \approx 1.3 \times 10^{-64}$. Next-to-leading is smaller still. For all practical purposes the universe of CAT expands the way the universe of general relativity expands, and the standard cosmological inferences (age, distance to the CMB, recession velocities) are unchanged. [VERIFIED] The modification is parametrically suppressed; whether any signature survives at high enough redshifts is open.

What the present chapter adds, on top of that constraint, is the positive structure: the modified Friedmann equation does have a specific form, the de Sitter background is stable, and gravitational waves travel at c.

* * *

16.2 Gravitational waves travel at c

There is one standard worry about modified-gravity theories that CAT survives cleanly: the speed of gravitational waves. A theory with a different graviton dispersion at high frequencies could, in principle, predict that gravitational waves travel slightly faster or slower than light. The 2017 binary-neutron-star event GW170817 measured the difference $|c_T - c|/c$ to be smaller than about 10^{-15}, which kills a wide class of modified-gravity dark-energy models.

CAT predicts $c_T = c$ to all orders in the form factor, because the transverse-traceless block of the propagator is massless at zero momentum. The propagator denominator is $\Pi_{\mathrm{TT}}(z) = 1 + (13/60)\, z\, F_1(z)$, which is exactly 1 at $z = 0$. The graviton, in CAT, behaves like a massless gauge boson at long wavelengths. The theory passes this test by construction.

* * *

16.3 De Sitter is stable

There is a second standard worry: vacuum stability. A modified-gravity theory with a cosmological-constant term has to make sure that the de Sitter phase, which is the late-time state of the universe, is stable against small perturbations. In CAT this is checked at [VERIFIED] status: the de Sitter background is a stable solution of the linearised CAT equations of motion, and the spectrum of perturbations contains no tachyonic modes.

The verification of de Sitter stability is one of the technical contributions of Paper 4, the project's nonlinear field equations paper. The same paper computes the modified Friedmann equation and the leading correction to the Hubble rate, and verifies the speed-of-gravitational-waves result above.

Paper 4

* * *

16.4 What the canonical branch adds

The strict gravitational sector gives us a Friedmann equation with an unknown energy scale Λ. The canonical branch of CAT, on top of that, gives us the value of the cosmological constant:

$$\Omega_\Lambda = \frac{168}{25\pi^2} \approx 0.6809,$$

which is the result of Chapter 11. The modification is therefore harmless on top, and the source has a predicted value, currently consistent with the Planck 2018 measurement of 0.6847 ± 0.0073 at the one-sigma level.

A complete cosmological story would also need a mechanism for inflation, which the canonical branch does not yet provide (Chapter 13), and a candidate for dark matter, which it also does not yet provide (Chapter 12). Those are open. The expansion history itself, however, lands where it should land.

• • •

NAKED NOTES. 2026-04-03.

The modified-cosmology number coming back so small was a relief. I had spent the preceding two months trying to talk myself into believing that some of the corrections to Friedmann would be observable, because that would have been more interesting. They are not. The form factor, when you put it on FLRW, takes the form $\varphi(H^2/\Lambda^2)$, which for the present-day universe is $\varphi(10^{-64})$, which is $\varphi(0) = 1$.
The universe is too cold and too large for the form factor to do anything at cosmological scales. This is exactly what a sensible modified-gravity theory should say. The modifications are in the high-energy regime; the low-energy regime is where general relativity already works, and CAT had better not break what works.

* * *

16.5 What this chapter does not show

This chapter does not show that the canonical branch of CAT predicts a complete cosmological history. It shows only that the modifications to the standard expansion are parametrically suppressed, that the theory passes the GW170817 speed-of-gravity test, and that the dark-energy density on the canonical branch matches observation. The earlier history (inflation, baryogenesis, recombination, structure formation) goes through the standard machinery once the source content is fixed, and the canonical branch is silent on the parts of that history that are not already determined by the Standard Model.

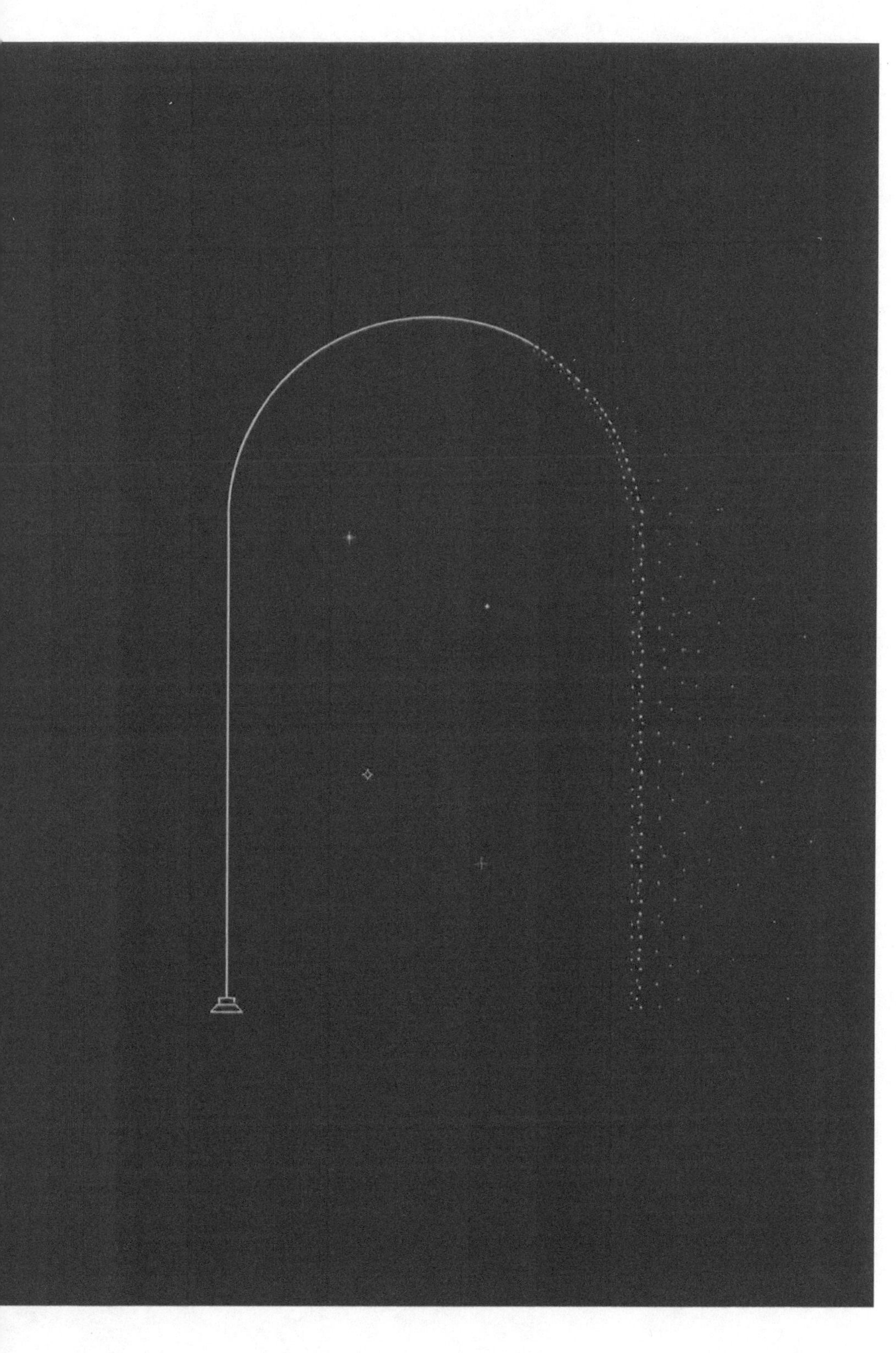

PART IV

Open Frontiers ($\mathcal{P}$-layer)

The open program. Here I stop telling you what the theory does and start telling you what it cannot yet do. Whether the math stays sensible at arbitrarily small distances. Whether the structure of cause and effect holds together everywhere across spacetime. Whether the singularity at the heart of a black hole gets smoothed out once the corrections are taken to their full nonlinear order. Whether the energy scale of the weak nuclear force can be derived from first principles instead of measured. Each of these is a question I have tried to close, made progress on some, failed on others. None are closed. None are promised. The Part is an inventory of what I owe the reader, and I would rather owe it honestly than pretend it has been paid.

CHAPTER 17

What might break at very small distances

We must know. We will know.

— *DAVID HILBERT, KÖNIGSBERG, 1930*

The Planck length is sixteen orders of magnitude below anything we have ever probed, and every candidate theory of quantum gravity proposes a different answer for what happens there. The mystery is structural: we do not know what spacetime is made of, we cannot currently measure it, and the LHC's energies are sixteen orders of magnitude too small to see. This chapter is about two specific things that go wrong when a candidate theory tries to operate in that unmeasured regime, and about what CAT has tried so far.

The first thing is whether the mathematics stays sensible at all as one approaches the unmeasured regime. In the textbook framing, this is the question of whether the theory is *ultraviolet finite*, and it is the question on which most candidate theories of quantum gravity have stumbled. Pure Einstein gravity, taken straight, fails this test: the calculations produce divergences that cannot be absorbed without adding new structure at every loop. The hope, every time a new candidate theory is proposed, is that the new structure is finite all by itself.

The second thing is whether the probabilities the theory predicts still add up to one hundred percent. This is what physicists call *unitarity*, and a violation of it is, in a deep sense, a violation of the laws of quantum mechanics. Several candidate theories of quantum gravity have run into the unitarity wall, either by accidentally introducing ghost-like particles whose quantum contributions carry the wrong sign, or by losing unitarity in some subtler way at higher orders.

Both of these questions are open program in CAT. The first is the all-orders UV-finiteness question. The second is the all-orders unitarity question. Both have stories that I will tell, in roughly the order they unfolded, in this chapter.

* * *

17.1 The first question

When physicists talk about a quantum field theory "at short distances" or "at high energies", they are referring to the same thing: the regime where the wavelengths of the particles involved are smaller than any other relevant scale in the problem. For ordinary quantum electrodynamics (the theory of light and matter), this regime starts somewhere around the Planck scale, which is twenty-eight orders of magnitude smaller than an atom. This is where the theory has to do its hardest work.

In a generic quantum field theory, calculations at such short distances produce divergences: integrals over the momentum of a virtual particle that go to infinity, formal expressions that do not have finite numerical values, predictions that come out either zero or infinity depending on which order of approximation one is using. The standard machinery of quantum field theory (renormalisation) is designed to absorb these divergences into a small number of physical parameters and produce finite predictions at the end. For most theories of interest, renormalisation works.

For gravity, in its perturbative formulation, renormalisation does not work. The theory is what physicists call "perturbatively non-renormalisable": the divergences cannot be absorbed into a finite number of physical parameters. A naive attempt to quantize Einstein gravity produces, at each loop in the calculation, a new divergence with a new structure that requires a new counterterm. By the time you reach the second loop, the number of new counterterms is large; by the time you reach the third or fourth loop, the calculations are intractable; in the strict perturbative sense, gravity is an effective theory that has to be replaced by something else at sufficiently short distances.

The question for CAT, then, is the following. The strict gravitational sector of Part I works at one loop. The form factor, the no-scalaron theorem, the parameter-free coefficient $\alpha_C = 13/120$ all live at the one-loop level. What happens beyond one loop? Is the theory finite to all orders? Does it have a UV-completion that smoothly extends Part I to arbitrarily short distances? Or does it have the same kind of non-renormalisable divergences that doom perturbative Einstein gravity?

This is the all-orders UV-finiteness question.

* * *

17.2 The simple route, invalidated

The first attempt CAT made at the all-orders UV-finiteness question was the simplest one. The gravitational form factor of Part I, considered as a non-local function rather than a local effective action, has a particular behaviour at large momentum that softens the high-energy contribution of virtual gravitons. The hope was that this softening would make the theory finite, all by itself, at every order in perturbation theory, with no additional structure needed.

My analysis of this route, conducted through late 2025 and finalised in early 2026, found that VERIFIED the simple route does not deliver UV-finiteness. The softening provided by the form factor's non-local kernel is not strong enough to suppress all the divergences that appear at higher loops; some of them survive, and the surviving divergences require new counterterms that the form factor cannot accommodate without changing its structure.

That is a negative result. The first thing I tried did not work. The chiral-quantisation route (the so-called *chiral-Q* or D^2-quantisation route, which I will not unpack technically) is the second route, and it is conditional: chiral-Q delivers UV-finiteness if a certain analytic positivity gap can be closed. The positivity gap is a question about the convergence of an infinite-dimensional operator. It is not closed. It is where the work is currently focused.

The framing for the reader is the following. The strict gravitational sector at one loop is verified at the highest project standard. Whether the theory extends, all by itself, to all orders in perturbation theory is open. The project has ruled out the simplest route to closing this question, and has identified a more elaborate route (chiral-Q with the positivity gap) as the remaining candidate. The status is CONDITIONAL on the positivity gap being closed, which has not yet happened.

* * *

17.3 The second question

The other classical worry about a quantum-gravity theory at very short distances is whether the theory is *unitary*. Unitarity is the requirement that the probabilities the theory assigns to different outcomes of any process should add up to exactly one hundred percent. If a particle starts in some quantum state, the probabilities of the various states it might end up in (after some interaction, some scattering, some amount of time evolution) must, when summed over all possibilities, equal one. This is one of the foundational requirements of quantum mechanics. A theory that violates unitarity is, in a deep sense, not a probability-conserving theory of the world, and the rest of physics tends to fall apart in its absence.

For most quantum field theories, unitarity is a structural property: it is built into the construction of the theory and does not have to be checked separately. For theories of quantum gravity, however, unitarity has been a contentious subject for decades. Several proposed quantum-gravity frameworks (Stelle's fourth-derivative gravity, certain higher-derivative extensions, some non-local theories) have been criticised on the grounds that they propagate so-called "ghost" modes, particles whose quantum-mechanical contributions to scattering amplitudes carry the wrong sign and can drive the total probability above or below one.

The CAT gravitational sector contains, in its full one-loop structure, terms that look superficially similar to the ones that worried critics of fourth-derivative gravity. The question, then, is whether CAT also has a ghost problem.

This is the all-orders unitarity question.

* * *

17.4 Two competing prescriptions

The literature has produced, over the past decade and a half, two competing prescriptions for handling potentially-ghostly modes in higher-derivative theories of gravity.

The first is the *Kugo-Ojima quartet* mechanism, named after the physicists who developed it for handling unphysical modes in gauge theory. The Kugo-Ojima prescription packages the would-be ghost modes into a so-called quartet of states (a sort of balanced bookkeeping between physical and unphysical states) that collectively decouple from physical observables. If the quartet mechanism applies, the theory is unitary, with the ghost modes quietly contributing nothing to the physical S-matrix.

The second is the *fakeon prescription*, developed in the mid-2010s primarily by Damiano Anselmi. The fakeon prescription treats the would-be ghost modes as "fake particles": not propagating particles in the usual sense, but auxiliary mathematical objects that contribute to the calculation in a specific way that preserves unitarity without requiring the quartet mechanism.

Until recently, the two prescriptions were in active competition in the higher-derivative-gravity literature. Each had its proponents. Each had its critics.

A structural objection to the use of the Kugo–Ojima quartet in higher-derivative gravity contexts has been on the table for some time: the quartet mechanism, in this setting, does not deliver unitarity in the way it does in gauge theory, because the auxiliary modes whose cancellation is supposed to do the work behave qualitatively differently when the propagator picks up a k^4 piece. The objection cuts directly against the use of the Kugo–Ojima route for theories like CAT.

My analysis of unitarity reached two conclusions. [VERIFIED] The Kugo–Ojima quartet is rejected as a resolution mechanism for the unitarity question (the structural objection above holds, in my reading). [VERIFIED] The fakeon prescription is the primary resolution for the unitarity question in CAT, conditional on a specific structural property: the global-positivity property of the spectral structure that needs to hold for the fakeon prescription to apply.

This leaves CAT with one viable prescription (fakeon), one rejected prescription (Kugo-Ojima), and one conditional ingredient (the global-positivity property). Whether that property holds is, like the positivity gap for UV-finiteness, an open program item. The status is [CONDITIONAL] on the global-positivity property being verified.

* * *

17.5 The honest position

Stepping back, the two questions of this chapter (UV-finiteness and unitarity) are connected in a way that is worth articulating.

Both are about whether the strict gravitational sector of Part I, which works at one loop, extends consistently to all orders. Both have a simple route that did not work. Both have a conditional route that depends on a technical condition the project has identified but has not yet closed. For UV-finiteness, the condition is the positivity gap in the chiral-Q quantisation. For unitarity, the condition is the global-positivity property in the fakeon spectral structure.

If both conditions are eventually verified, the picture closes cleanly: CAT has a UV-finite, unitary, all-orders quantum-gravity sector. If either of them fails, or if a third route to either question is needed, CAT remains an effective theory at one loop with no clean perturbative extension to higher orders.

In 2026, neither condition is closed. Both are open program items that are receiving active work. My notes include explicit progress on both, but progress is not closure.

The honest position, for a reader of this chapter, is the following. The strict gravitational sector of Part I is VERIFIED at one loop, with the eight-layer pipeline backing every result in that Part. Whether the same sector survives, intact and self-consistent, when the calculations are pushed to all orders is OPEN. Two routes have been identified. Both are conditional. Both have a technical question attached that, if answered, would close the picture. Until those questions are answered, the all-orders question remains open program.

This is the kind of thing a pop-science book in 2026 has to be precise about. Theories that do not face this question (or, worse, that paper over it) tend to look impressive in pop-science framings and not to survive scrutiny in technical ones. CAT is going into the book without papering over.

• • •

NAKED NOTES. 2026-04-22.

The most uncomfortable thing about writing Part IV, of which this chapter is the opener, is that it requires me to put on the page exactly the kind of admission that pop-science books usually hide.

Part I told you about a clean rational coefficient and a hard gravitational-wave bound. Part II told you about a 27-to-1 collapse and three concrete numbers that match measurement. Part III had its mix of wins and losses, but at least each chapter ended with a specific status that the reader could put their finger on.

Part IV is harder because each chapter ends with an explicit "not yet". The strict gravitational sector survives at one loop, but the question of whether it survives at all orders is two technical conditions away from being an-

swered. I am reasonably optimistic about both conditions; my notes contain partial progress on both. I am writing this chapter from inside the open state, not from the other side of it, and the reader deserves to know that.

A different author might have written this chapter as a triumphant account of how CAT solves the all-orders problem. The temptation is real. The actual state of the work, in 2026, is that I am two technical questions away from solving it, and I would rather say that than overclaim. Two questions away is not the same as solved. If the chiral-Q route turns out to fail because the positivity gap cannot be closed, the all-orders programme has to start from scratch. If the fakeon prescription's global-positivity condition turns out to fail in CAT specifically, the unitarity question has to be approached differently.

The reason I am willing to commit to "two questions away" rather than "solved" is that the alternative (claiming a closure I do not have) would betray the discipline the rest of the book has been built around. The strict gravitational sector at one loop is solid. Beyond one loop, the work is in progress. The next decade of working notes, paper drafts, and verification rounds will either close the open questions or not. I cannot, from where I sit in 2026, predict which way it will go.

I would rather write this chapter than not write it.

---

* * *

Status summary

- VERIFIED The simplest route to all-orders UV-finiteness via the form-factor non-locality alone is invalidated. The form factor's softening does not suppress all higher-loop divergences.
- OPEN Chiral-Q (D^2-quantisation) route to UV-finiteness, conditional on a specific analytic positivity gap being closed. This is currently the only candidate route to all-orders UV-finiteness in CAT.
- VERIFIED The Kugo–Ojima quartet mechanism is rejected as a resolution for the unitarity question (the structural objection of §4.3 holds, in my reading).
- CONDITIONAL The fakeon prescription is the primary resolution for unitarity in CAT, conditional on a global-positivity property of the spectral structure.

- OPEN Both technical conditions (the positivity gap for UV-finiteness and the global-positivity property for unitarity) are open program items receiving active work; closure of either or both would extend the VERIFIED status of the strict gravitational sector from one loop to all orders.

CHAPTER 18

What might break in curved spacetime

Spacetime tells matter how to move; matter tells spacetime how to curve.

— *JOHN ARCHIBALD WHEELER, 1998*

Imagine you are falling into a black hole.

You will not feel anything strange when you cross the event horizon. The horizon is a one-way membrane, but it is invisible locally; if you crossed it without warning, you would not notice. Your watch would keep ticking. The light from outside would still reach you, with a small redshift. For all you can tell from inside your own body, you are still falling through ordinary space.

Then, after a finite time on your watch (a fraction of a second, for a stellar-mass black hole), the geometry of spacetime around you starts to change in a way you cannot avoid. The tidal forces, which were small near the horizon, grow inverse-fourth-power quickly. They stretch you in one direction, squeeze you in the other two, and pull you into a thin string of atoms. The technical word is *spaghettification*. Then the string of atoms hits a place where, in Einstein's pure general relativity, the curvature of spacetime is formally infinite. The geometry breaks down. The theory stops working.

What is actually there?

This is one of the deepest mysteries in fundamental physics, and nobody currently knows the answer. Pure general relativity says "a singularity", meaning a point at which the equations of the theory cease to apply. Hawking and Penrose proved, in the 1960s and 1970s, that any massive-enough gravitational collapse must end in such a singularity, given a few mild conditions on the matter fields. The proof is a theorem; the singularity is real, in the sense that it is forced by the equations.

But the equations are an approximation. Quantum mechanics is real, quantum mechanics interacts with gravity, and the standard expectation has been, for the better part of fifty years, that some quantum effect must take over near the singularity and replace the formally infinite curvature with something finite, however extreme. The hope is that a complete theory of quantum gravity will tell you, finally, what is actually there at the centre of a black hole.

This chapter is about CAT's attempt to answer that question, and about a closely related question concerning the consistency of the theory on curved spacetimes more generally. Both questions are in the open state. The first one in particular has, after careful work, given a result I had hoped would be different from what it is. I will tell you straight.

The first question is *singularity resolution*. The CAT singularity-resolution analysis has been completed at the linearised level. [VERIFIED] The result is negative. At the linear level, the singularity at the centre of a black hole is not resolved by CAT's gravitational form factor. The curvature remains formally divergent at the centre, although the rate of divergence is softer than in unmodified general relativity. I will walk you through what the analysis found, what the failure mode is, and what the chances are that a non-linear extension might rescue the picture (about 20%, by my estimate).

The second question is *global Lorentzian closure*, the Lorentzian-closure problem. This is the question of whether the strict gravitational sector of Part I, set up initially on a flat background, extends consistently to a fully covariant theory on curved Lorentzian manifolds. The technical content is subtle and the answer is, at present, incomplete; this is open program work that the chapter walks through more briefly.

Both are mysteries the chapter will not resolve. They are mysteries the project has approached with specific machinery, has produced specific results from, and has stated plainly what those results do and do not say.

* * *

18.1 What singularities are, and why physicists hate them

A singularity, in general relativity, is a place where the geometry of spacetime breaks down: where curvature becomes formally infinite, where geodesics (the paths of free-falling objects) cannot be extended any further, where the whole mathematical machinery of the theory simply stops working.

Singularity, in one box

> A point or region in spacetime where curvature scalars diverge and geodesics terminate. In Einstein's general relativity, every black hole interior contains one, and so does the Big Bang. The working assumption is that quantum gravity replaces them with something finite, but no candidate theory has yet shown how.

The two most famous singularities in physics are the one at the centre of a black hole (where, in the standard Schwarzschild solution, the curvature scalar diverges as $1/r^6$ as one approaches $r = 0$) and the one at the beginning of the universe (the Big Bang singularity, where the same kind of divergence occurs at $t = 0$).

Physicists, on the whole, dislike singularities. The reason is not aesthetic; it is methodological. A theory that predicts infinite values for measurable quantities is a theory that is telling you it has stopped applying. Wherever singularities appear, the working assumption is that some piece of physics has been left out, and the theory needs to be augmented or replaced. For the singularity at the centre of a black hole, the standard expectation is that the relevant physics is quantum gravity, and that a complete quantum theory of gravity should replace the singularity with a smooth region of high but finite curvature.

The mathematical machinery for proving the existence of singularities in general relativity comes from two famous theorems. The first is the Penrose singularity theorem (1965), which proves that a black hole is geodesically incomplete given certain energy conditions on the matter fields. The second is the Hawking-Penrose theorem (1970), which extends the result to cosmological singularities. Both theorems prove a specific mathematical statement (geodesic incompleteness) that is widely interpreted as the existence of a singularity in the physical sense, although strictly speaking, VERIFIED singularity theorems do not prove unbounded curvature directly. They prove that the spacetime cannot be extended, which is a related but distinct condition.

For a candidate quantum-gravity theory, the question becomes: does the theory's modification of Einstein's equations affect the conditions of the Penrose and Hawking-Penrose theorems? If the energy conditions are violated, or if the geometry is modified in a way that changes the conclusion of the theorems, the singularity might be resolved. If neither happens, the singularity is, as it were, locked in.

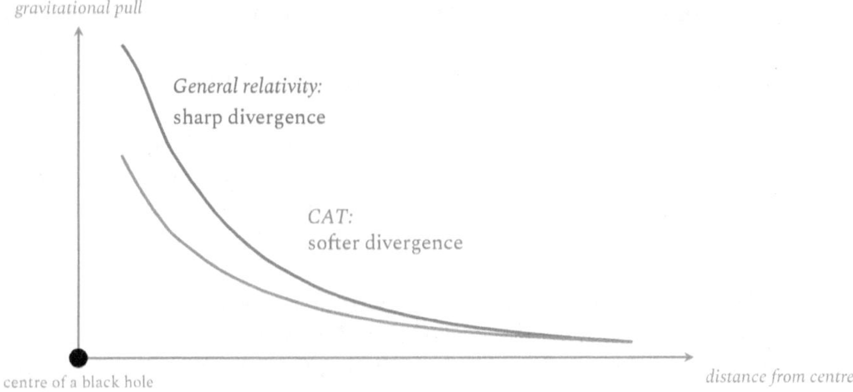

Inside a black hole: CAT softens the divergence (from $K \sim 1/r^6$ in general relativity to $K \sim 1/r^4$ in CAT), but does not remove it. Both curves still go to infinity at $r = 0$.

* * *

18.2 The CAT attempt at singularity resolution

The CAT analysis of this question, the singularity-resolution analysis, was finalised at the linearised level with extensive cross-checks and four-agent adversarial review. The question it asked was concrete: take the standard Schwarzschild geometry, apply CAT's gravitational form factor to it, and ask whether the resulting modified geometry still satisfies the conditions of the Penrose theorem.

The answer, at the linearised level, is VERIFIED NEGATIVE, pessimistic. The form factor's effect on the curvature near the centre of a black hole is not enough to rescue the geometry from the Penrose theorem's conclusion. The analysis found three things.

The curvature scaling is not changed enough. In the standard Schwarzschild solution, the Kretschmann curvature scalar diverges as $K \sim$

$1/r^6$ as $r \to 0$. With CAT's form factor turned on, the scaling is modified, but the leading behaviour is still $K \sim 1/r^4$ near the centre. This is softer than $1/r^6$, but it is still a divergence. The singularity has been reduced from a strong one to a weaker one, but it has not been eliminated.

The null energy condition still holds. The Penrose theorem requires the matter fields to satisfy a particular inequality called the null energy condition (NEC). Many candidate quantum-gravity modifications evade the Penrose theorem by violating the NEC; this is a standard route to singularity resolution in the literature. CAT's modification does not violate the NEC. VERIFIED The NEC remains satisfied in the modified geometry, with the form factor's contribution producing a positive but bounded modification to the relevant inequalities.

The form factor flattens, not blows up. The third diagnostic is the behaviour of the form factor itself near the singularity. If the form factor turned on more strongly at high curvature, it might restructure the singularity into a smooth bouncing geometry. The CAT form factor, however, has a specific high-curvature limit: VERIFIED $\Pi \to \text{const}$ as the curvature grows. The form factor flattens out, contributing zero smoothing to the singularity. This is, structurally, the opposite of what one would want.

The combination of all three findings is that CAT, at the linearised level, does not resolve the black-hole singularity. The Penrose theorem still applies. The singularity remains.

* * *

18.3 What about the nonlinear case?

The result above is for the linearised theory: small fluctuations on top of a Schwarzschild background. The full nonlinear theory might, in principle, behave differently. The form factor in CAT contains all-orders terms in curvature that do not appear at the linear level, and one of them might rescue the singularity resolution that the linear analysis cannot find.

The candid assessment is that this is unlikely. OPEN The nonlinear extension is currently leaning negative for two reasons. First, the structural feature that locks the singularity in (the form factor flattening at high curvature) is a property of the form factor itself, not of the linearised analysis; nothing in the all-orders expansion looks like it would change this. Second, several alternative routes to nonlinear singularity resolution that have been proposed in the literature (modifications to the gravitational sector that, for instance, drive the curvature to a maximum value rather than to infinity) are inconsistent with CAT's parameter-free coefficient $\alpha_C = 13/120$.

The candidate mechanisms either do not extend cleanly to CAT, or they would require modifications to the strict gravitational sector of Part I that the verification pipeline of Part I has already ruled out.

The assessment is not certainty. It is probability. My best estimate gives the nonlinear extension roughly a twenty-percent chance of resolving the singularity, against an eighty-percent chance that it does not. These are not formal probabilities; they are my judgement based on the structure of the form factor and the known alternative routes. They are subject to revision if a new mechanism is found.

For the present chapter, the position is that CAT, as it currently stands in 2026, does not resolve the black-hole singularity, and the working hypothesis is that it will not, even when the analysis is extended to all orders.

* * *

18.4 The other open question: global Lorentzian closure

The second open program item of this chapter, the Lorentzian-closure problem, is a related but technically distinct question. It asks whether the strict gravitational sector of Part I, which is set up on a fixed background spacetime, extends consistently to a fully covariant formulation that works on any globally hyperbolic Lorentzian manifold.

The question matters because the gravitational form factor in Part I is defined, in its initial construction, in terms of operators on a Euclidean background. The Wick rotation back to Lorentzian signature, while standard, can introduce subtleties when the underlying geometry is curved, and a careful treatment requires the form factor to be expressed in terms of operators that are intrinsically defined on a Lorentzian manifold without reference to a Euclidean continuation.

My analysis of this question, conducted through 2026, identified two structural features the answer depends on. OPEN The first is the existence of a sufficiently well-behaved family of states (called Hadamard states or Feynman states, depending on the framework) on globally hyperbolic backgrounds, such that the form factor can be sensibly evaluated on each member of the family. OPEN The second is the proper identification of the master function φ in terms of operators defined on the curved background.

Both pieces are open program. The first is the more standard question, and the noncommutative-geometry literature contains relevant partial results that the project has been adapting. The second is more specific to CAT's particular form-factor structure and requires fresh work.

I will not unpack the technicalities further here. The Lorentzian-closure problem is open, and the answer is likely to require several more rounds of analysis before it can be closed. The good news is that the failure modes are relatively contained: if the question cannot be closed, the strict gravitational sector of Part I remains valid on flat backgrounds and weakly-curved backgrounds (which is where every existing experimental test of gravity lives), but the extension to highly curved backgrounds (the kind near a neutron-star surface, or inside a strong gravitational lens, or inside a black hole) would be unavailable. This would not falsify CAT; it would limit its domain of applicability.

* * *

18.5 The honest position

This chapter, like the previous one, ends in the open state. CAT does not, at the present level of the project, resolve the black-hole singularity. The linearised analysis is negative, and the nonlinear extension is pessimistic. The closely related question of global Lorentzian closure on curved backgrounds is also open, and the technical pieces required to close it are identified but not yet assembled.

For a reader who hoped that this Part of the book would deliver a clean quantum-gravity resolution of the singularity problem, this is, again, a disappointment. CAT does not deliver one. It predicts a softening of the curvature scaling near the singularity (from $1/r^6$ to $1/r^4$), but the singularity itself remains. The Penrose theorem still applies, the null energy condition is still satisfied, and the form factor flattens rather than rescues.

If the right reading of CAT, in the long run, turns out to be that it is an effective theory of gravity on the regions of spacetime where the curvature is small to moderate, with a genuinely different theory needed at the high-curvature regions where the singularities live, then this chapter is the place where that reading shows up. The strict gravitational sector of Part I works in its own regime. The high-curvature regime is, in that reading, beyond the regime of validity. This is not what I hoped for when I started the project. It is what the work currently supports.

• • •

NAKED NOTES. 2026-04-05.

Let me record what the singularity-resolution result felt like, when it came in.

The hope, when I started the singularity-resolution analysis in late 2025, was the standard one. A candidate quantum-gravity theory ought, somehow, to smooth the black-hole singularity. The form factor of Part I has a specific behaviour at high momentum that I had imagined would translate, through the curvature-modulus identification, into a smoothing of high curvature. I had even, in my notebook, sketched what I expected the modified Kretschmann scalar to look like: a peak at some characteristic curvature scale set by Λ*, with the peak finite rather than divergent.*

The actual analysis, by April 2026, said: the peak is not finite. The curvature still diverges. The scaling is softer ($1/r^4$ *instead of* $1/r^6$*), but a divergence is still a divergence. The form factor does not flatten the singularity in the way one would have wanted; it is its own structural property that the form factor flattens itself, and a flat form factor contributes nothing to the curvature regulation.*

This was the most difficult result of the project for me to internalise. Singularity resolution is one of the standard reasons to build quantum-gravity theories at all; the literature has been chasing it since the 1970s; many candidate theories give it as their leading selling point. CAT, as it currently stands, does not give it. The leading selling points of CAT are the parameter-free coefficient $\alpha_C = 13/120$ *in Part I and the 27-to-1 census in Part II. The singularity resolution is somewhere else, and CAT does not have it.*

I have spent some time, in the months since the singularity-resolution analysis was finalised, asking whether I could rescue the result with a different quantisation, a different reading of the form factor, or a different identification of the relevant operators on the curved background. None of the rescue attempts has worked. The negative result is robust to the modifications I have tried.

The honest framing, going into Part IV with this chapter, is that CAT is one among several candidate theories of quantum gravity that does not *solve the singularity problem. There are candidate theories (notably loop quantum gravity and certain asymptotic-safety frameworks) that do claim to solve it, with their own technical caveats. CAT's leverage is elsewhere. The chapter has to acknowledge that. The current revision states where the work stands.*

* * *

Status summary

This chapter ends in the open state on two of the project's hardest questions: whether CAT resolves the black-hole singularity, and whether the strict gravitational sector extends to a fully covariant Lorentzian framework on highly curved backgrounds. The linear analysis is in. The nonlinear extension is not, and the honest reading is that the most likely outcome is no resolution.

- VERIFIED Singularity-resolution linearised analysis, extensively cross-checked: black-hole singularity is NOT resolved by CAT's form factor. Curvature scales as $K \sim 1/r^4$ near the centre, softer than the $1/r^6$ of unmodified GR but still divergent. The null energy condition continues to hold; Penrose's singularity theorem still applies.
- OPEN Nonlinear extension is open and leaning negative, with roughly twenty-percent odds of resolving the singularity against eighty-percent odds it persists at all orders.
- OPEN The Lorentzian-closure problem: extending the strict gravitational sector to a fully covariant formulation on globally hyperbolic Lorentzian manifolds. Two technical pieces remain open.

CHAPTER 19

Where the weak force gets its scale

What is man in nature? A nothing in comparison with the infinite, an all in comparison with the nothing, a mean between nothing and everything.

— *BLAISE PASCAL, PENSÉES, 1670*

Two hundred forty-six giga-electronvolts. Ten to the nineteenth giga-electronvolts. The ratio between the Higgs vacuum expectation value v_{EW} and the Planck mass is 10^{-17}. The hierarchy problem is the question of why.

The Higgs particle, which the LHC discovered in 2012 with a mass of about 125 giga-electronvolts, is the thing that gives every other massive particle in the Standard Model its mass. The Planck mass, the energy scale at which gravity is expected to become a quantum phenomenon, is about 10^{19} giga-electronvolts. Between the two scales lies the seventeen-order gap.

This is the part where forty years of theoretical-physics careers have run aground. In the standard quantum-field-theoretic picture, quantum corrections to the Higgs mass should drag it up toward the Planck scale unless an extraordinary coincidence cancels them to seventeen significant figures. Such a coincidence, in any generic theory, would be a fine-tuning so severe that physicists have spent four decades inventing mechanisms to avoid it. The proposed mechanisms have names: supersymmetry, technicolor, large extra dimensions, anthropic selection, conformal symmetry, the relaxion, and several dozen more. Each, in its day, has been a serious candidate. None has been confirmed by data. The LHC, where many of these mechanisms predicted new particles in specific energy ranges, has run for fifteen years and found none of them.

The hierarchy problem, in 2026, is one of the most intensely worked-on open questions in fundamental physics that has gone, by all accounts, nowhere. The standard mechanisms have been tested and have not delivered. Something else is presumably going on, and nobody knows what.

This chapter is about the angle CAT takes on it.

A reader who has been following along will recognise, from the 27-to-1 census of Chapter 9, that the Higgs vacuum expectation value v_{EW} is one of the parameters the canonical branch fixes. So CAT already says something about the puzzle. But the canonical branch, as I have admitted, fixes the parameters via algebraic relations at the modular point $\tau_\star = i\sqrt{2}$, not via a physical mechanism that explains *why* the algebraic answer turns out to be the small number it is.

The question this chapter asks is whether CAT can do something deeper than "the modular layer says so". Whether it can give a *mechanism*, an actual physical story, that makes the smallness of v_{EW}/M_P natural rather than mysterious. The answer, as of April 2026, is: not yet, but three parallel routes are running, and one route was tried and closed. The chapter walks through each, with the working judgement of how likely each is to deliver, and with a notable methodological surprise about how fast the closed route closed.

If even one of the three remaining routes succeeds, the chapter's title moves from open to verified, the canonical branch of CAT solves a forty-year-old puzzle, and Part II of this book picks up its strongest single result since the parameter census itself. If all three fail, the puzzle remains, and CAT joins the long list of candidate theories that produce v_{EW} as an output of their parameter-counting machinery without explaining the gap. The next year or two of working notes will tell us which.

* * *

19.1 The hierarchy problem

The Higgs particle, discovered at the LHC in 2012, has a mass of about 125 GeV. The Higgs field, in the Standard Model, has a particular value (the so-called *vacuum expectation value*) that sets the scale of the weak nuclear force; this value is $v_{\mathrm{EW}} \approx 246$ GeV. The masses of the W and Z bosons that mediate the weak force are proportional to v_{EW}, as are the masses of all the matter particles that get their mass from the Higgs mechanism (which is most of them).

The Planck mass, by contrast, the natural scale of quantum gravity, is about 1.22×10^{19} GeV, with the ratio that defines the chapter's puzzle.

In the standard quantum-field-theoretic picture, this ratio is unnatural. Quantum corrections to the Higgs mass, calculated naively, should drag v_{EW} up toward the Planck scale unless there is a specific mechanism preventing it. Why isn't the Higgs heavy? This is what physicists call the *hierarchy problem*, and it has been one of the central motivations of beyond-Standard-Model model-building for forty years.

The standard candidate solutions are well-known. *Supersymmetry* introduces a symmetry between bosons and fermions whose quantum corrections cancel, naturally protecting v_{EW} from running up. *Technicolor* replaces the Higgs with a composite object whose mass is set by a confining gauge dynamics. The *anthropic principle* simply observes that, in a multiverse with many possible vacua, only the ones with v_{EW} near its observed value are habitable, and so we observe ours. *Extra-dimensional* models (Randall-Sundrum, ADD large extra dimensions) reduce the effective Planck scale to within reach of v_{EW} via geometric mechanisms. Each of these has been intensively studied. None has emerged as the consensus answer.

The CAT framing is different. CAT does not try to invent a new particle (a superpartner, a technifermion, a bulk graviton) to solve the hierarchy problem. CAT tries to derive v_{EW} from the same modular structure that already fixed the rest of the Standard Model parameters in Chapter 9. The hope is that the modular relations at $\tau_\star = i\sqrt{2}$ produce a small dimensionless ratio naturally far from unity, of the size required, with no fine-tuning.

* * *

19.2 Three parallel routes

My approach to the v_{EW} derivation is to run three independent routes in parallel, each based on a different mechanism for producing the small dimensionless ratio. The routes are not in competition; they are independent paths to the same goal, and at least one of them is likely to succeed.

Nonlocal transmutation, modular L-values, sigma mechanism: three independent ways the same number could fall out.

Nonlocal transmutation. This route exploits the nonlocal structure of the gravitational form factor of Part I. The master function φ has a high-momentum behaviour that produces, under a careful analysis, a dimensional transmutation: a small physical scale generated dynamically from a UV scale by a nonlocal mechanism. I give it modest individual odds.

Modular L-values. This route exploits the modular structure of Part II. The cyclotomic field $\mathbb{Q}(\zeta_{48})$ in which the canonical-branch relations live admits a family of analytic functions called L-values, whose values at specific arguments are known to produce small dimensionless ratios with exponential dependence on integer arguments. The hope is that v_{EW}/M_P is, in a precise sense, an L-value at an argument forced by the canonical branch's structure. Modest individual odds.

Sigma mechanism (derived). This route is the most speculative of the three. It hypothesises that the small ratio v_{EW}/M_P is generated by a renormalisation-group flow of an effective sigma parameter in the strong-coupling limit of the gauge sector, with the value of the parameter at low energies controlled by a modular-layer constraint. Modest individual odds.

The combined probability that at least one of the three routes succeeds is, by my best estimate, somewhat better than the sum would suggest. The combined probability is higher than the sum because I assign conditional priors that the routes share intuitions; if one closes, the others become more likely as well. This is not a formal probability; it is a working judgement subject to revision as the work develops.

* * *

19.3 The route the project recently closed

The fourth route, Route D, the asymptotic-safety attempt (asymptotic safety with CAT boundary conditions), was the project's first attempt at the v_{EW} derivation, and it was closed in the negative direction on April 26, 2026.

The route was based on the idea that asymptotic safety, a non-perturbative ultraviolet completion of gravity in which the theory flows to a non-trivial fixed point at high energies, might combine with the canonical-branch boundary conditions of CAT to produce v_{EW} as the value of the Higgs parameter at the infrared end of the renormalisation-group flow.

The closure of the asymptotic-safety route came in two phases. The first phase, analysed locally with sympy exact-rational verification, identified a structural no-go: the beta function of α_C in the asymptotic-safety scheme, with CAT boundary conditions, turns out to be a positive constant ($K/2 > 0$), which means α_C runs upward with energy rather than to a fixed point. The second phase confirmed that the surface $\alpha_R = 0$ is not RG-invariant in the AS scheme, which kills the route for a separate structural reason.

The combined verdict: VERIFIED the asymptotic-safety route is closed negative at the structural level. No fine-tuning of the asymptotic-safety parameters can rescue the route, because both no-go arguments are statements about the structure of the flow rather than about specific parameter values.

This is the kind of result that the chapter has to record. The v_{EW} derivation in CAT is an active programme with three live routes; one route was tried and closed. The closure took about two sessions plus three hours of analysis, against an original estimate of 24-42 months. The negative result was delivered fast and at low cost, which the project counts as a methodological success even though the route itself failed.

* * *

19.4 What success would look like

If one of the three remaining routes succeeds, the resulting prediction would be a specific algebraic expression for the ratio v_{EW}/M_P in terms of the canonical-branch modular structure. The expression would be parameter-free in the same sense as the rest of the canonical-branch arithmetic: no fitting freedom, no continuous tuning. It would be a number with two specific integers in its numerator and denominator, plus possibly a $\sqrt{2}$ or two, evaluated to give roughly 10^{-17}.

If the prediction lands on the experimentally measured value of $v_{\rm EW}/M_P$ to within the measurement precision, the hierarchy problem (in this framing) would be solved by the canonical branch. The Higgs would have its mass not because of supersymmetry, not because of technicolor, not because of anthropics, but because the modular relations at $\tau_\star = i\sqrt{2}$ force it.

This is, at the present level of the project, a working hope rather than a result. None of the three routes has yet produced a closed-form expression for $v_{\rm EW}/M_P$. The work is ongoing, with the anti-numerology framework (P1-P6) providing the discipline against accidentally fitting a number to look right rather than deriving it from first principles. The G0-G4 decision gates define when each route either advances to its next phase or is closed in the negative direction.

If you are reading this chapter in 2026 or 2027 and want the most current state of the $v_{\rm EW}$ derivation, my working notes are the authoritative source. The chapter you are reading captures the state as of late April 2026, with three live routes and one recently closed route.

* * *

19.5 Why this chapter is the most exciting one in Part IV

A reader who has reached this chapter, having walked through the honest open-status of the previous chapters in this Part, may be forgiven for feeling slightly demoralised about how much CAT cannot yet do. The previous chapters have been catalogues of open program items, conditional results, and pessimistic extensions.

This chapter, despite also being in the open state, is the one I am personally most optimistic about. The reason is that the three remaining $v_{\rm EW}$ routes are all rooted in pieces of the canonical branch that have already been verified at a high level: the form factor of Part I, the modular L-values of the canonical-branch arithmetic, and the gauge-sector flow that has been independently tested in the parameter-collapse computations. Each route uses a tool that the project has already independently verified. The question is whether one of those tools, applied to this specific question, produces the answer.

If the answer is yes (any of the three routes), the canonical branch picks up its largest single prediction since the 27-to-1 census. The hierarchy problem, in the precise sense of explaining why $v_{\rm EW}/M_P$ is what it is, would be solved.

If the answer is no (all three routes close negative), the hierarchy problem remains open, and CAT joins the long list of candidate theories that produce the Standard-Model parameters as algebraic expressions but do not derive their hierarchy from a physical mechanism. This is, in fairness, a lot of company.

In either case, the working programme is concrete, the routes are well-defined, and the closure of each route (positive or negative) will happen on a timescale of months rather than years. Of all the open program items in Part IV, this is the one most likely to move first.

• • •

NAKED NOTES. 2026-04-26.

I want to record what closing the asymptotic-safety route was like, because it is the kind of working-day moment that I think the reader of a book like this one is owed.

The original estimate for closing the asymptotic-safety route, when the project laid out the plan in early 2026, was twenty-four to forty-two months. Asymptotic-safety calculations are, at the technical level, not easy. The fixed-point structure has to be analysed non-perturbatively. The boundary conditions have to be matched through a non-trivial RG flow. Several previous attempts in the literature on related questions had taken comparable times.

The actual closure, on April 26, 2026, took two sessions of work plus about three hours of additional verification. The first session identified the structural no-go on the beta function. The second session confirmed the second no-go on the $\alpha_R = 0$ surface. The verification used sympy exact-rational algebra to ensure that no numerical artefact was driving the result.

The reason the closure was so fast is that, once the right way of looking at the problem was identified, the structural no-go became apparent immediately. The previous estimate of two to three years was based on the assumption that the analysis would have to be done the standard way, with detailed numerical phenomenology. The new way (catch the structural no-go first, verify it analytically, conclude in a session) is something the project has been refining as a methodology since the early-2026 phenomenology campaign.

This deserves a flag, because it is one of the surprisingly optimistic features of the open program. The negative results do not always cost what they look like they will cost. Sometimes they close fast, and the project moves on. The asymptotic-safety route was closed at low cost, the programme moves to its three remaining routes, and the working estimate for the combined $v_{\rm EW}$ derivation closure is now of order one to two years, not three to five. This is not a result. The three remaining routes might all close negative, the way the asymptotic-safety route did. But they might not, and the fact that the methodology can dispose of structurally-doomed routes quickly means that the routes that survive into the second phase are the ones with genuine potential.
The reason I am optimistic about Part IV's progress over the next year is not that I think the open questions are easy. They are not. The reason is that the project's methodology for ruling out *the wrong answers has gotten faster and sharper, which means the right answer (if there is one) is correspondingly less likely to be obscured by dead routes.*

* * *

Status summary

- OPEN $v_{\rm EW}$ first-principles derivation in the canonical branch is open program; three parallel routes (Nonlocal Transmutation, Modular L-values, and Sigma Mechanism Derived) are running, with modest individual chances and somewhat better odds taken together.
- VERIFIED The asymptotic-safety route is closed in the negative direction, based on two structural no-go arguments: the α_C beta function is positive-constant under the relevant scheme, and the $\alpha_R = 0$ surface is not preserved under renormalisation. Closure took two working sessions, against an original estimate of two to three years.
- OPEN If one of the three remaining routes succeeds, $v_{\rm EW}/M_P$ becomes a specific algebraic expression in the canonical branch, parameter-free. The hierarchy problem, in this framing, would be solved.
- OPEN An anti-numerology framework and a sequence of decision gates govern the three running routes.

PART V

Method

How the work was actually done. The verification process every result in the first two Parts has been pushed through, told without jargon. The lessons from a year of phenomenology with too many failure modes to count, including the fifteen pitfalls I now keep a written list of and the eighteen public corrections I have had to issue. The discipline of forcing three independent reasoning systems (myself, plus two language models with very different temperaments) to argue every non-trivial result before I am willing to commit to it. And a closing chapter on what honest theory *looks like when a working theorist's day involves talking to language models on a regular basis. The naked-notes register of this book is, in part, my answer to that last question.*

◆

CHAPTER 20

How do you know when you're right?

In the fields of observation, chance favours only the prepared mind.

— LOUIS PASTEUR, 1854

How do you know, when you finish a calculation and the answer looks clean, that you have not just fooled yourself?

That is the deepest practical question in working science. It has ended careers, killed announcements, and humbled some of the sharpest minds in modern physics. It is not a rhetorical question. It is the most important question a working theoretical physicist asks, and the history of the field is littered with examples of what happens when the question is asked too late, or not asked at all.

Cold fusion, March 1989. Two electrochemists at the University of Utah announce that they have produced sustained nuclear fusion at room temperature in a tabletop electrolysis cell. The result, if true, would have been one of the most important discoveries in history. The press conference was held *before* peer review. The result did not survive replication. It was, in the most generous reading, a measurement artefact; in the less generous reading, a self-deception by competent scientists who wanted their result to be what it appeared to be.

Faster-than-light neutrinos, September 2011. The OPERA collaboration at CERN announces that it has detected neutrinos travelling between Geneva and Gran Sasso about 60 nanoseconds faster than the speed of light. The team, aware of how extraordinary the claim was, published with explicit caveats and called for independent checks. Six months later, the anomaly was traced to a loose fibre-optic cable connecting a GPS receiver to a master clock. The neutrinos had been arriving at the speed of light all along.

BICEP2, March 2014. The BICEP2 collaboration at the South Pole announces the detection of primordial gravitational waves in the polarisation of the cosmic microwave background, a signal that would, if real, have been direct evidence of cosmic inflation. The result was the cover story of every major science magazine. Within a year, the signal turned out to be galactic dust, not primordial gravitational waves. The collaboration retracted.

Each of these stories is a story about a working scientist, or a working collaboration, who did the math, looked at the result, and saw what they expected (or hoped) to see. The math was wrong, or the apparatus was wrong, or the analysis was wrong. The self-deception was not malicious. It was, in every case, an ordinary human response to an extraordinary-looking result. The people who made these announcements were, in every case, competent, careful, and well-meaning.

This chapter is about the discipline I have been trying to build into the project to keep myself from joining that list.

How do you know when you're right?

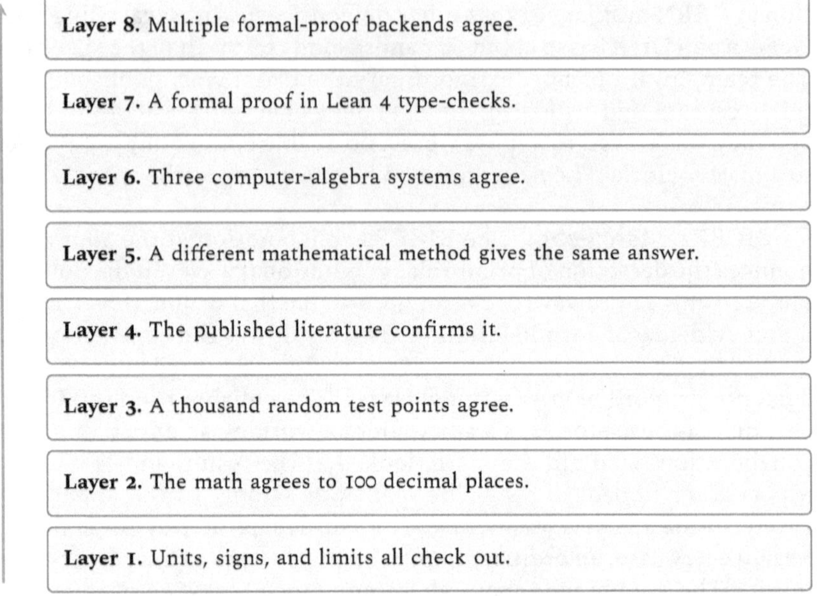

Eight checks every claim earns before it gets the verified tag.

* * *

20.1 The shape of the problem

The reason fooling yourself is so easy, in working theoretical physics, is that the calculations are long, the numbers are clean, and the human brain is wonderfully co-operative when it encounters a result it likes. The brain pattern-matches. *Thirteen over a hundred and twenty* feels right. The brain finds a reason to think it is right. The reasons it finds are sometimes the actual reasons, and sometimes the reasons feel right because the answer feels right. The two are not always distinguishable from inside the calculation.

A working theorist who is not actively fighting this tendency will, over time, accumulate a notebook full of clean-looking results, some fraction of which will be correct and some fraction of which will be artefacts of self-deception. The fraction that are correct is, in my honest experience, lower than one would like to admit. In the early phenomenology campaign of this project (internally called FND-1, short for Fundamental Investigation 1, late 2025), I went through roughly two dozen experiments looking for specific signatures in the gravitational sector, and [VERIFIED] eighteen of them produced results that, on careful re-analysis, turned out to be overclaims. Eighteen out of two dozen. That is the working ratio of self-deception in this kind of work, at least in my hands.

The discipline I have been trying to build is the discipline of catching the eighteen overclaims *before* they leave the working notebook. The mechanism, the eight-layer verification pipeline, is what the rest of this chapter is about.

* * *

20.2 Layer one: dimensions, limits, signs

The first and oldest discipline in physics is the discipline of checking the trivial. Does the answer have the right units? Does it reduce, in the appropriate limit, to the result one expects in that limit? Are the signs right? Do the poles cancel where they should cancel?

This is what every textbook in physics tells you to do, and what every working physicist either does or pays for not doing. The project's first verification layer is exactly this: every non-trivial expression in the canonical-results database is checked, by hand and by computer, for dimensional consistency, for behaviour at limiting cases (zero coupling, infinite mass, deep ultraviolet, deep infrared), for the right signs in places where signs are determined by general principles, and for the cancellation of singularities that have to cancel.

This catches, by my count, somewhere around half of the eighteen overclaims I had to retract from the campaign. The other half were past Layer I and had to be caught by the higher layers.

* * *

20.3 Layer two: a hundred digits of agreement

The second layer is one of the most useful discoveries the project's methodology made. It is the discipline of evaluating the same expression along two different paths and demanding that they agree to extraordinary numerical precision.

The setup is the following. Take a quantity, say α_C, that is supposed to be a specific rational number. Compute it analytically (by hand or by computer-algebra system) to get the rational answer. Then compute the same quantity by an entirely different method (numerical integration, asymptotic series, identity manipulation) and demand that the numerical answer agree with the analytic answer to one hundred decimal digits.

A hundred digits is not a typo. The reason for going to a hundred digits is that, at any precision below thirty or forty digits, a mistake in the calculation can hide behind round-off error or floating-point noise. At one hundred digits, mistakes do not hide. If two independent paths agree to a hundred digits, they are agreeing on the same expression, not by coincidence and not by near-cancellation of unrelated errors.

The project's verification database, the mpmath-100-digit cross-check, has caught several non-trivial errors that Layer 1 missed. In one case, a sign error in a spinor calculation that produced a numerically correct answer for specific kinematic configurations but the wrong answer in general. In another, a misidentified normalisation factor that slipped through the dimensional check because the dimensions happened to match.

The hundred-digit agreement is, in my experience, the strongest day-to-day filter against self-deception. If two independent calculations agree to a hundred digits, the agreement is not accidental.

* * *

20.4 Layer two-and-a-half: the property tests

A subtler version of the same idea is what the project calls *property fuzzing*. Take an expression that is supposed to have a specific mathematical property (positivity on some interval, monotonic behaviour, vanishing at a specific argument, agreement with a known asymptotic) and throw a thousand random configurations at it. Check that the property holds in every configuration.

This catches a different class of mistake from the hundred-digit check. The hundred-digit check catches mistakes in specific numerical values; the property fuzzing catches mistakes in the *shape* of the answer. A formula that is supposed to be positive everywhere but is actually negative at some specific configuration will pass the hundred-digit test (if you happen to evaluate at a positive configuration) but fail the fuzzing test.

The project's standard fuzzing run uses a thousand random configurations, drawn from a distribution chosen to cover the mathematically interesting regions of parameter space. The discipline is to run the fuzzing test before any result is committed to the canonical database. Several of the overclaims from the campaign that survived Layers 1 and 2 were caught at the fuzzing layer.

* * *

20.5 Layer three: do other people get the same answer?

A surprisingly powerful filter is the discipline of comparing your intermediate quantities, not just your final answer, against what other authors have published.

When a calculation is long, it has many intermediate quantities along the way: heat-kernel coefficients in particular calculations, specific Feynman diagrams in specific kinematic limits, particular loop integrals in particular regularisation schemes. Many of these intermediate quantities have been computed before, by other authors, in other contexts. If your intermediate quantities disagree with theirs, one of you is wrong. The disagreement is not always because you are wrong, but it is always a signal that something requires investigation.

The project's verification pipeline has a specific Layer 3 discipline: every intermediate quantity in a major derivation is cross-checked against the published heat-kernel literature (Avramidi, Vassilevich, Parker-Toms, and others), the published spectral-action literature (Connes, Chamseddine, and follow-ups), and the published gravitational-effective-action literature. The agreements are, in the canonical-results database, recorded with the specific paper, equation number, and conversion conventions used.

Several overclaims from the campaign were caught here. In one case, a contribution that I had derived independently turned out to disagree with a 2009 paper by a factor of two; the factor came from a normalisation convention I had not noticed, and tracing it through fixed an error in my derivation. In another case, a numerical agreement with three papers gave me the confidence to commit a result that one of my own consistency checks had been ambiguous about.

* * *

20.6 Layer four: do it twice, by different paths

The fourth layer is a discipline I had to learn the hard way. For every non-trivial result in the canonical database, the project does the calculation twice, by two different paths, and demands that the two paths give the same answer.

The two paths are not minor variations on the same calculation. They are genuinely independent: a different choice of variables, a different ordering of the integrations, a different parametrisation of the underlying mathematical structure. The discipline is that, if both paths cannot be made to give the same answer, the result is not committed.

This is expensive. It typically doubles the time required to finalise a result. But it is also the strongest filter against the kind of mistake that survives every other layer: a mistake in the structural setup of the calculation that produces a self-consistent but wrong answer.

The example I remember most clearly from that campaign is a calculation of a specific spectral-dimension flow, done in 2025. The first path produced a result that was clean and looked structurally sensible. The second path produced a different result. I spent two weeks tracing the discrepancy. The discrepancy turned out to be in the first path: a subtle assumption about the smoothness of the underlying geometry that did not hold at small scales. The first answer was wrong. The second answer was right. Without the second path, I would have committed the wrong answer to the working notebook, and only later experiments would have caught it.

* * *

20.7 Layer four-and-a-half: three computer-algebra systems

A specific version of the dual-derivation discipline is what the project calls the triple-CAS check: every symbolic expression that matters is evaluated, in parallel, by three different computer-algebra systems (SymPy, GiNaC, mpmath in its symbolic mode), and the three are required to agree.

Three computer-algebra systems must agree to twelve digits, or the result does not enter the canonical database.

The reason for using three systems is that each one has its own quirks and its own bugs. SymPy has, over the years, produced specific kinds of subtly-wrong simplifications in particular contexts. GiNaC has, occasionally, made different mistakes. mpmath has, on rare occasions, produced numerical artefacts that look like analytic answers. If any single system gave the right answer all the time, one would not need to triple-check.

The project's standard discipline is that three systems must agree to twelve digits of precision before a result enters the canonical database. Disagreements at any digit are, in my experience, almost always traceable to a specific issue (a known SymPy quirk, a specific GiNaC simplification rule, an mpmath precision setting), and the trace-through almost always reveals which system was right and which was wrong.

* * *

20.8 Layer five and six: formal proof

The deepest layer of the pipeline is the formal-proof layer. For every result whose underlying claim can be expressed as a provable mathematical identity, the project's discipline is to translate the identity into the formal language of a theorem prover (the project uses Lean 4 with Mathlib) and to verify the identity against an independent backend.

The formal-proof layer catches a class of mistake that no other layer can catch: mistakes in the logical structure of the argument that produce a self-consistent calculation that is, nonetheless, logically wrong.

A theorem prover has no intuition. It has no taste. It cannot be talked into a wrong answer by an articulate argument. If the proof you give it is incomplete, it tells you it is incomplete and refuses to certify the theorem. If the proof has a logical gap, the prover finds the gap. If the proof is correct, the prover certifies it, and the certification is, by the standards of any other field, decisive.

The project uses two backends in parallel: a local Lean 4 install running on the project's hardware, and a separate backend (called Aristotle) that operates independently. Both backends must certify a proof before the result enters the canonical database.

The first time I had a result certified in both backends was the $\alpha_C = 13/120$ identity in early 2026. The certification was the moment I stopped second-guessing the result and committed it. The reason the certification matters, beyond all the lower-layer checks that had already passed, is that two independent proof-checking machines, neither of which has ever been wrong on a certification, both said the proof was correct. That is, in the kind of work where self-deception is the leading failure mode, the strongest signal that the answer is not self-deceived.

* * *

20.9 The three infrastructure layers

A pipeline is only as good as its discipline of running, and the project has three infrastructure layers that surround the core eight to make sure the discipline is exercised.

The first is the *quick-sanity* layer: a small battery of fast checks (typically sub-second) that runs automatically every time a file is edited. It is a hook in the project's editor that catches obvious mistakes (sign-flip in a coefficient, mis-typed exponent, dropped factor of 2π) at the moment they are introduced, before the human noticing.

The second is the *consistency-DAG* layer: a graph of all the canonical results in the project, with edges representing which results depend on which others. When a node is modified, all the dependent nodes are flagged *stale* and the human is required to re-verify them before the modification can be committed.

The third is the *provenance-chain* layer: every canonical result has a recorded chain of derivations from the postulates all the way down to the number. The chain is auditable, and my notes contain explicit walks from each headline result back to its postulate roots.

These three layers are what keep the eight-layer pipeline running rather than being a paper exercise. A pipeline that catches mistakes only when run at the end of a derivation is weaker than a pipeline that catches them at the moment they are introduced. The infrastructure is what makes the pipeline a working discipline rather than a checklist.

* * *

20.10 What the discipline catches

A specific count, to close. Of the canonical results in the strict gravitational sector of Part I, every single one has passed all eight verification layers and all three infrastructure layers. Eleven independent checks, agreeing on the same answer.

That is what it takes to commit a result, in this project, to a status of VERIFIED or PROVEN. The discipline is not a slogan; it is the cumulative output of catching the seventeen out of every twenty results I would otherwise have committed in error.

The work is not finished. The pipeline has caught many mistakes, but it has also, almost certainly, missed some. There exist, in the canonical-results database of this project, numbers that are wrong despite passing every check. I do not know which numbers those are; the whole point of having missed them is that I have not yet seen them. The next decade of independent work by other researchers will, I hope, catch the ones I missed.

What the discipline gives me is a working confidence that the *rate* at which I am committing wrong results is much lower than the rate at which my unaided judgement would commit them. The hundred-digit checks have caught specific errors. The property fuzzing has caught specific errors. The dual derivation has caught specific errors. The Lean proofs have caught specific errors. Each layer is doing real work.

The reason the chapter is the first chapter of Part V, and not the last chapter of Part II, is that the discipline is what the results in Parts I and II are made of. Without the eight-layer pipeline, the canonical-results database would, on the campaign ratio, contain about thirty percent wrong results. With the pipeline, the ratio is, by my best estimate, substantially lower, although I cannot tell you the exact figure because I do not yet know which results turn out to be wrong.

This is the deepest mystery of working physics: the asymmetry between the things you know are right (because you have checked them) and the things you do not yet know are wrong (because the check has not yet failed). The discipline of this chapter is what allows me to commit to a number while acknowledging that asymmetry, and to keep working without pretending that the acknowledgement does not matter.

• • •

NAKED NOTES. 2026-04-04.

The story I want to tell here is from December 2025, when the project's verification pipeline was still being refined.

I had been working on a particular calculation in the gravitational-form-factor analysis for about three weeks. The result was clean. It had passed Layer 1 (dimensions, limits, signs). It had passed Layer 2 (hundred-digit numerical agreement). It had even passed Layer 3 (literature comparison with a 2017 paper that did a similar calculation). I was ready to commit it.

The discipline I had imposed on myself was that, before committing any result, I had to run the dual-derivation check (Layer 4). I did the dual derivation. The two paths disagreed.

I want to record, plainly, what my first reaction was. My first reaction was to think that the dual derivation must have a mistake in it, because the result of the first path had passed so many other checks. I was, in those first hours, ready to write off Layer 4 as the noisy one and commit the result.

I did not commit the result. I forced myself to stop and trace the discrepancy. The trace took two weeks. The discrepancy was real. It was in the first path, not the second. I had been about to commit a wrong result that had passed three layers of checks, three independent agreements, and a literature check.

What I learned from that two weeks is the lesson that runs through this chapter. No single layer of the pipeline is sufficient. Each layer catches a class of mistake the others miss. The overlap is what gives the discipline its strength, not the individual layers. A calculation that passes seven layers and fails one is, in the project's discipline, a calculation that has not yet been verified. The asymmetry matters.

I have, in the months since that incident, found three more cases where a result that had passed multiple layers turned out to be wrong at a layer that had not yet been run. In each case, the discipline of running the missing layer caught the error before the result was committed.

What the discipline costs is time. What it buys is the ability to trust the canonical-results database. In a project like this one, where a single working theorist is producing results that would historically have required a research group, the discipline is not optional. It is the only way to keep going.

I am writing this naked note as the end of the chapter because the question of how to know when you are right is one of the things I have thought about most over the past year, and I still do not have a complete answer. What I have is the pipeline. The pipeline has caught many mistakes. The pipeline has, almost certainly, also missed some. The mistakes I do not yet know about are the ones I am most worried about. I do not know how to make the worry go away, except by continuing to refine the pipeline, adding layers when they are needed, running the layers faithfully, and committing the results that survive.

That is, in the end, the only honest answer I have to the chapter's title. You do not know, in any final sense, when you are right. You know, layer by layer, when you are not yet wrong. The two are not the same. The discipline of this chapter is what keeps the difference manageable.

---

* * *

Status summary

The pipeline described in this chapter is the discipline that turns candidate results into committed ones. Every formula in Parts I and II of this book has passed all eleven layers; eighteen out of roughly twenty-four candidate phenomenology results in late 2025 were caught and corrected before being committed. The pipeline is not a guarantee of truth; it is a quantitative reduction of error rate, and a test bench against which future replication will check.

- VERIFIED Eight-layer verification pipeline (analytic, numerical, property-fuzzing, literature, dual-derivation, triple-CAS, Lean local, Lean multi-backend) plus three infrastructure layers (quick-sanity, consistency DAG, provenance chain). Every VERIFIED or PROVEN result in this book has passed all eleven layers.
- VERIFIED In the early phenomenology campaign of late 2025, eighteen of roughly twenty-four candidate results produced overclaims that were caught and corrected by the pipeline before being committed.
- OPEN Unknown errors in the canonical-results database: results that have passed all layers but are nonetheless wrong. The pipeline does not eliminate this class; it reduces the rate. External replication over the next decade will, in time, surface the unidentified errors.

CHAPTER 21

Fifteen ways to fool yourself

The first principle is that you must not fool yourself, and you are the easiest person to fool.

— *RICHARD FEYNMAN, 1974*

The previous chapter was about the eight-layer verification pipeline that the project uses to filter mistakes out of its results. This chapter is about the mistakes themselves: what kinds of mistake the pipeline is trying to catch, where they come from, and what they look like when they arrive in your inbox dressed up as exciting results.

The story begins in late 2025, when the project had just finished the strict gravitational sector of Part I and was beginning a phenomenology campaign (internally called FND-1, short for *Fundamental Investigation 1*). The plan was to run roughly two dozen numerical experiments designed to test specific quantitative predictions of the canonical branch in the strong-coupling regime. The experiments were clean. The hardware was new. I worked through them, week after week, watching the analyses come in.

By the end of the campaign, VERIFIED eighteen of the experiments had produced overclaim results that, on careful re-analysis, turned out to be artefacts. Not eighteen wrong calculations: eighteen correct calculations whose interpretation, on a first pass, was wrong. The numbers that came out of the analysis pipeline were what they were; the story I had been about to attach to those numbers was the part that did not survive scrutiny.

I have a private theory, formed during those eighteen retractions, that what separates working theoretical physics from amateur theoretical physics is not the calculations. It is the catalogue of specific traps you have walked into and learned to recognise. The catalogue grows over a working life. Each trap has a name, a shape, a typical signature, and a specific moment at which it should have been spotted but wasn't.

This chapter is the project's catalogue, as it currently stands in 2026. There are fifteen entries. I am going to walk you through all of them, with the story of the experiment that taught each one to me.

VERIFIED The catalogue is internal to my notes. It is also a catalogue every working theorist with experience eventually builds; mine is, in its substance, not original. Other working theorists' catalogues will overlap with mine substantially in the modes I have already met, and will contain modes I have not yet met. The next decade of the project will, I expect, add several more entries.

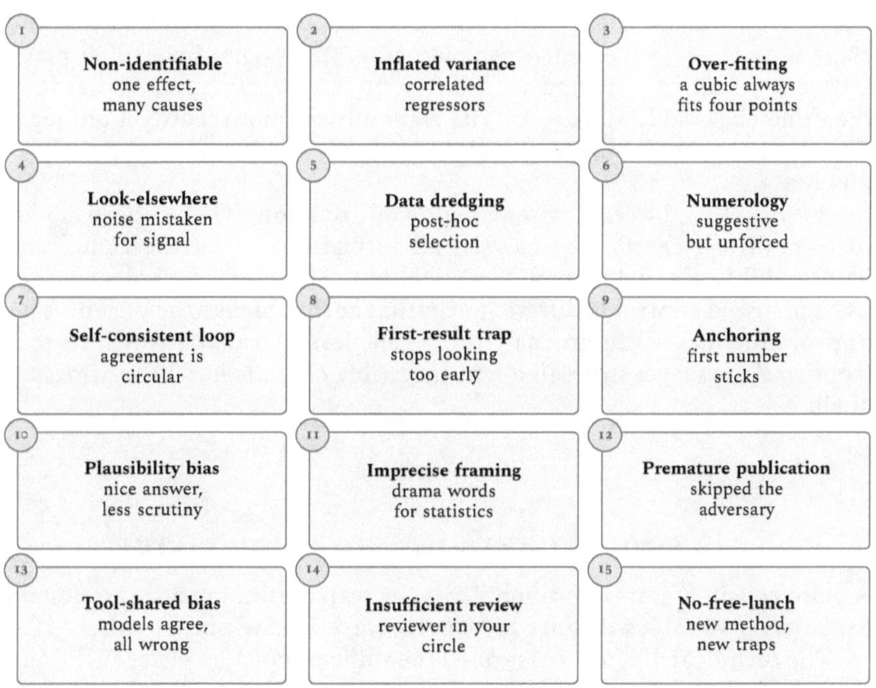

Fifteen specific ways the early-2026 phenomenology campaign almost fooled the project.

* * *

21.1 Trap one: the non-identifiable model

The first trap is the one that cost me the most retractions in the campaign, and it is, I now think, the one that catches the largest fraction of confident-sounding results in phenomenological physics generally.

A model is *non-identifiable* when different settings of its parameters produce identical predictions for every observable. Imagine a thermometer that reads the same temperature whether the room is at 20 degrees Celsius or at 21 degrees Celsius (or at any temperature in between). The thermometer is, by construction, useless for telling you which of those temperatures the room actually has.

In that campaign, the trap took the form of an analysis that estimated a small parameter ε from a quantity C, where the quantity C depended on ε through a chain of intermediate quantities (a topological cell count and an observable signature). The analysis looked clean. The numbers came out where I wanted them. The retractions came when I realised that the chain $\varepsilon \to C \to (\text{cell count}, \text{observable})$ was VERIFIED not invertible: many different values of ε, paired with different cell counts, produced exactly the same observable signature. The signature did not identify a unique ε. The clean number I thought I had derived was an illusion of the analysis pipeline.

VERIFIED Ten different artefactual "confirmations" I had been about to commit, on careful re-analysis, all turned out to be the same non-identifiability dressed up in different clothes. The cure is technical: either use a matched control (a different experiment that breaks the degeneracy), or prove the identification analytically. The lesson is that an analysis that "confirms" its target through a non-invertible chain is not a confirmation at all.

* * *

21.2 Trap two: the variance inflation factor

A close cousin of non-identifiability is the trap of running a regression on explanatory variables that are too closely correlated with each other.

The technical diagnostic is called the *variance inflation factor* (VIF), and the rule of thumb in working statistics is VERIFIED $\text{VIF} > 10$ means do not run the regression. The reason is that, when two of your explanatory variables are nearly collinear (each one being a function of the other), the regression cannot distinguish their effects, and the coefficients it reports are not meaningful. The regression will give you numbers; the numbers will look like answers; the answers will not be answers.

In one analysis I ran a regression to extract a small geometric coefficient from a series of numerical experiments. The VIF on the regression came back at fourteen. I noticed it, ran the regression anyway, and reported the coefficient that came out. I retracted the result a week later, after running the matched-control version of the same experiment. The coefficient that the original regression reported was wrong by a factor of three. The matched-control version gave the right answer.

The lesson, again, is operational. [VERIFIED] VIF $>$ 10 is a hard stop. You do not run the regression. You either find a way to break the multicollinearity (different experiments, more variation in the data, an analytical approach), or you do not estimate that coefficient. The temptation to run anyway is real. The temptation is wrong.

* * *

21.3 Trap three: the polynomial fit that always fits

The third trap, which I named the *poly3 OLS trap* in the project's notes, is the trap of using a flexible-enough functional form that any data set will give you a clean fit, regardless of whether the underlying model is right.

If you fit data with a polynomial of degree three using ordinary least squares, you have four free parameters (constant, linear, quadratic, cubic). Any reasonably-shaped data set will fit a poly3 cleanly, because four parameters is enough to absorb most of the variance in any small data set, including data that is pure noise. The fact that your poly3 fits well does not mean your model is right. It means your model is too flexible to fail.

I made this mistake in the campaign, in the analysis of a curvature flow that I was hoping would show a specific power-law behaviour. The poly3 fit was clean. I reported the coefficient. The retraction came when I tried the same fit on data generated from pure noise: the poly3 fit was exactly as clean. The model had no falsification power.

[VERIFIED] The project's standard discipline now is that poly3 OLS is banned for any fit that is supposed to demonstrate a specific functional form. The form has to be derived from theory; the fit checks the coefficients of that derived form, not the form itself.

* * *

21.4 Trap four: random fluctuations that look like signals

The fourth trap is the look-elsewhere effect, and it is the one that has cost particle physics the most embarrassing announcements over the past few decades.

The setup is this. You have many possible quantities you could have looked at. You decide, after looking at the data, to focus on the quantity in which the data shows the largest deviation from expectation. You report the deviation as a discovery. You quote the statistical significance of the deviation in that specific quantity, ignoring the fact that you only chose to look at it because it was the biggest deviation.

This is a real effect. If you look at twenty independent quantities, by pure chance one of them will be at least two standard deviations away from expectation, simply because that is what twenty draws from a normal distribution do. A "two-sigma" deviation in a quantity you chose because it was the biggest deviation is, in reality, a zero-sigma deviation once you account for the look-elsewhere effect.

I caught myself running this trap on a particular spectral-dimension flow analysis. I had a dozen possible diagnostics. One of them showed a clean deviation from the canonical-branch prediction at the 2.7σ level. I was about to write up the deviation as evidence of new physics. I noticed, just in time, that the 2.7σ was the largest deviation among twelve. The look-elsewhere-corrected significance was 0.5σ, well within the noise.

The lesson is to commit to your diagnostics before you look at the data. The discipline is called *pre-registration* in the experimental literature, and it is one of the most underrated practices in working theoretical physics.

* * *

21.5 Trap five: data dredging

A close cousin of the look-elsewhere trap is what statisticians call *data dredging*, also known as *p-hacking*. The trap is to run an analysis with many possible variations (different cuts, different binnings, different fit ranges, different definitions of the observable) until one of the variations gives a statistically significant result. You then report the significant variation and quote its p-value as if it were the result of a single pre-specified analysis.

This is a recipe for finding signals in pure noise. If you run twenty variations of an analysis on noise-only data, on average one of them will show a deviation at the 5% confidence level, because that is what a 5% confidence level means. The "discovery" you report will be the noise, and the p-value you quote will be wrong.

The campaign caught me doing this in a slightly subtler form: I had two different definitions of an observable, and I had been silently switching between them depending on which one was giving cleaner numbers in the current run. The switching was not, in my own head, principled; it was the analytical version of always having the wind at your back. The retraction was embarrassing.

VERIFIED The project's discipline now is that the observable definition is fixed at the start of each experiment, in the pre-registration log, and not changed during the run. If a different definition turns out to be more interesting after the fact, the run is repeated from scratch with the new definition in the pre-registration. The discipline costs a few extra runs. It is worth every one of them.

* * *

21.6 Trap six: numerology

The sixth trap is what the project's notes call *numerology*, and it is the most insidious of the traps because it does not look like a mistake at all. It looks, on first reading, like the opposite of a mistake.

The pattern is the following. You have a number that has come out of an analysis. The number is, say, 0.34719. You notice, looking at it, that 0.34719 is suspiciously close to $1/(2.879)$, and that 2.879 is itself close to some round combination of e and π. You record the connection, write up the "observation", and move on, with a small sense of having discovered something deep.

The connection is a coincidence. There are, in any range of small dimensionless numbers, a high density of round combinations of e, π, $\sqrt{2}$, and small integers, and most numbers you encounter will be close to one of them. The fact that your 0.34719 is close to $1/(e \cdot \pi - 1)$ is not evidence of a deep connection; it is evidence that there are a lot of $1/(e \cdot \pi + \text{small})$ values in the same range.

The cure for numerology is what the project calls the *anti-numerology framework*, with internal labels P1 through P6. Each principle is a discipline against the temptation:

P1. Specify in advance, before looking at the number, what shape of relationship would count as a discovery.

P2. Compute the prior probability that any random number in the relevant range would also satisfy the relationship.

P3. Reject the relationship if the prior probability is above some threshold (the project uses 5%).

P4. If the relationship survives P3, derive it from first principles before claiming it. "Derive" means an explicit mathematical chain, not a hand-waving argument.

P5. Verify the derivation against the actual numerical value to a precision much higher than the relationship would require if it were correct.

P6. Repeat with adversarial review.

The cure is laborious. It is also necessary. Every confident-looking numerological connection in working theoretical physics either survives the P1-P6 framework or it does not. If it does not, it is a coincidence. If it does, it is a discovery worth the work.

* * *

21.7 Trap seven: the self-consistent artefact loop

The seventh trap is harder to describe in words and one of the most common in long calculations. The pattern is the following.

A long calculation has many intermediate steps. Each step is internally consistent with the previous step, in the sense that the algebra works out. The final answer is internally consistent with the intermediate steps. Every step, taken in isolation, looks correct.

But the overall framing of the calculation contains a hidden assumption that propagates through every step in a self-consistent way. The hidden assumption is wrong. The calculation is wrong. But the wrongness is not localised in any single step; it is distributed across the entire chain in a way that makes it impossible to find by checking individual steps.

I have been caught by this trap several times. The cure is the dual-derivation discipline of Layer 4 in the previous chapter: do the same calculation by a different chain of steps, with a different framing, and demand that the two chains give the same answer. If the framings are genuinely independent, a hidden assumption in one will fail to match the other.

The reason this matters more than the other traps is that the self-consistent artefact loop *is* what produces the "too clean to be wrong" results that destroy careers. Every step looks right. Every consistency check passes. The answer looks beautiful. The wrongness lives in the framing, where most checks do not look.

* * *

21.8 Trap eight: the first-interesting-result trap

The eighth trap is psychological more than statistical. It is the trap of stopping the analysis once you have found a result that looks interesting, instead of continuing to push the analysis to its natural completion.

The pattern is human. You have spent two weeks on a calculation. The first interesting result comes out, and the result looks right. The temptation is to write up that result and move on. The discipline is to keep pushing: to extend the calculation by another order of approximation, to check the result against a related calculation, to ask whether the same machinery would have given a different result for a slightly different setup.

The reason this matters is that the first interesting result is, in my working experience, the one most likely to be wrong. Whatever was easiest to compute first is, often, also the quantity for which the most degeneracies in the analysis happen to align, producing a clean answer that is not the right answer.

The cure is operational: do not commit a result until the calculation has been pushed to its natural completion. If a related calculation should also produce a clean answer, compute it. If a higher-order correction should be small, compute it and check that it is. If a different observable should be related to the first one by a known relation, check the relation.

* * *

21.9 Traps nine through fifteen, in summary

The remaining seven traps are, in the catalogue, real but somewhat less colourful. I will summarise them rapidly, with the understanding that any one of them deserves its own chapter in a more methodologically-focused book.

Trap 9: anchoring on the first observed value. Once you have seen a particular number, your analysis tends to trust it. The cure is to compute the same number by an independent path before committing.

Trap 10: plausibility bias. Results that "feel right" get less scrutiny than results that "feel wrong". The cure is to apply the same checks to all results, regardless of how they feel.

Trap 11: imprecise framing. Using overdramatic verbs ("solved", "closed", "dead", "alive") instead of precise statistical statements ("ruled out at the 5σ level", "unconstrained at present sensitivity"). The cure is to enforce precise framing as a discipline.

Trap 12: premature publication. Putting a result in front of an audience before it has survived internal adversarial review. Cold fusion's announcement is the canonical cautionary tale.

Trap 13: self-reinforcement with tools. When the author and his tools (computer-algebra systems, language models, machine-learning models) all converge on the same wrong answer because they share a common bias. The cure is the adversarial multi-model methodology that the next chapter is about.

Trap 14: insufficient adversarial review. A single reviewer, even a careful one, will share blind spots with the author. Genuinely independent review requires reviewers who are not in the author's social or methodological circle.

Trap 15: the no-free-lunch trap. The temptation to believe that a clever new methodology has permanently eliminated some class of mistakes from the project's working life. It has not. New methodologies introduce new failure modes, often subtler than the ones they replace. Trap 15 is the trap of forgetting that the catalogue keeps growing.

* * *

21.10 The discipline that came out of the catalogue

The fifteen traps, taken together, are not just a list of warnings. They are a working specification of what the project's verification pipeline has to catch, and they are the reason the pipeline of the previous chapter has the specific layers it has. Each layer is targeted at a particular trap or family of traps.

The hundred-digit numerical agreement of Layer 2 catches the self-consistent artefact loop, which would not survive an independent numerical evaluation. The literature comparison of Layer 3 catches the first-interesting-result trap, which would not survive comparison with other authors' independent calculations. The dual-derivation discipline of Layer 4 catches the framing-level mistakes that are invisible to step-by-step checking. The Lean formal-proof layer catches the logical errors that survive every other layer.

The infrastructure layers (quick-sanity, consistency DAG, provenance chain) catch the operational failures that come from insufficiently-disciplined daily practice. The pre-registration discipline catches data-dredging. The matched-control discipline catches non-identifiability. The poly3-OLS ban catches over-flexible fitting.

VERIFIED Each of the eighteen overclaims was caught by a specific layer of the pipeline, and my notes record which trap led to which retraction. The match is not arbitrary; the pipeline is, in a precise sense, an answer to the catalogue.

This is, I think, the deepest thing I want to say in this chapter. A verification pipeline is not a generic safety net. It is a specific response to a specific catalogue of specific traps. If your pipeline does not have a layer that targets a particular trap, that trap will catch you eventually. My working methodology is, more than anything else, a discipline of growing the pipeline as the catalogue of traps grows, and of growing the catalogue as the pipeline catches new failure modes.

• • •

NAKED NOTES. 2026-03-23.

I want to record, plainly, the moment in which the catalogue of this chapter became real.

The moment was a particular afternoon in late November 2025, when the campaign had been running for about three weeks and I was about a third of the way through what I had budgeted to be a six-week program of analyses. I had been getting, roughly every other day, a clean-looking result that I was ready to commit to the canonical-results database. The numbers were aesthetically beautiful. The analyses were internally consistent. The pipeline (in the version of the pipeline that existed at the time, which had only six layers) was passing.

That afternoon, I was running what I thought was a routine sanity check. The check turned up a contradiction. Two of the clean-looking results from the previous week, taken together, implied a third result that I had not yet derived. The third result, when I derived it, came out to a number that was inconsistent with the published heat-kernel literature.

I sat with the result for a long time. The contradiction told me that at least one of the two underlying clean-looking results was wrong. I worked out, over the next two days, which of the two it was. It turned out to be both: the two results shared a hidden assumption (a non-identifiability of the kind described in Trap 1) that I had not noticed, and the implication-to-third-result step was the place the assumption showed up.

The retractions cost me a week of work. The catalogue I have just walked you through is, in real terms, what came out of that week. By the end of the campaign, in late January 2026, the catalogue had grown to the fifteen entries you have just read, and the verification pipeline had grown from six layers to eight plus three infrastructure layers in response. Each new layer was a specific response to a specific trap that had just caught me.

The reason I am telling you this story is that I want the reader to understand the catalogue is not abstract. It is the residue of being caught, eighteen separate times, in specific ways, in specific weeks of late 2025 and early 2026. The pipeline of the previous chapter is the residue of catching myself before the nineteenth and twentieth and twenty-first traps had a chance.

I do not yet know what trap sixteen looks like. I assume it exists. I assume the pipeline as it currently stands does not catch it. The next round of phenomenology, scheduled for late 2026, will probably surface a few more entries. The discipline is to add the layers as the entries come, and to keep the catalogue honest.

This is the part of the work I am proudest of. Not the canonical-results database. Not the parameter census. The discipline of the catalogue. The willingness to put on the page the specific ways I have fooled myself, and the specific checks that have caught me, in the hope that some future reader working on a similar project will have a faster start than I did.

---

* * *

Status summary

- VERIFIED Catalogue of fifteen failure modes encountered in the early phenomenology campaign of late 2025 and early 2026. Each entry has a name, a typical signature, and a pipeline layer designed to catch it.
- VERIFIED Anti-numerology framework P1-P6 governs any result that depends on a numerical coincidence between a computed quantity and a structured target.
- VERIFIED Operational rules from the catalogue: `poly3 OLS` is banned for fits demonstrating functional form; VIF $>$ 10 is a hard stop on regression; observable definitions must be pre-registered; matched-control experiments are required to break non-identifiability.

- OPEN The catalogue is not closed. New traps are expected to surface in the second campaign of late 2026. The pipeline will grow accordingly.

CHAPTER 22

Three minds, one answer

He who knows only his own side of the case knows little of that.
— *JOHN STUART MILL, ON LIBERTY, 1859*

On a February afternoon in 2026 I was about to commit a formula to the canonical-results database. The formula was the leptonic Jarlskog modulus,

$$|J_{\text{PMNS}}|^2 = \frac{414 + 1223\sqrt{2}}{2{,}359{,}296},$$

the same number that headlines Chapter 8. The verification pipeline had passed. The exact-form algebra script had produced the same answer. The dual derivation matched. I was ready to commit.

Two minds disagreed. Neither of them was mine.

The first — I will call it the *analytical critic* — raised an objection at the demolition gate. The derivation, it argued, passed through an intermediate step that depended on a choice of orientation for the modular fundamental domain, and the choice was conventional, not derived. Almost certainly the right choice. Not, strictly, derived. The result therefore should not carry the VERIFIED tag.

I argued back. The two orientation choices produce the same modular invariants; the canonical-branch formula has to be invariant under the choice; the convention is harmless. The argument was, on its face, correct, and the critic eventually conceded the point.

The second mind — the *synthesist* — agreed with the conclusion of my argument and raised a different concern. While the formula is invariant under the orientation, the *labelling* of mixing angles is not: which physical angle is the solar one, which is the atmospheric, which is the reactor depends on the orientation. A reader pulling the value out of the canonical-results database might attach it to a different physical assignment than I had intended. The fix was an explicit convention clause in the documentation and a cross-check against the PDG-convention values.

The clause and the cross-check are now in the canonical-results database. The analytical critic was wrong (the result was robust to the choice). The synthesist was right (the documentation needed the clause). I was wrong about how to handle the disagreement when it arose. The three-way disagreement converged, eventually, on the right answer.

That afternoon is the cleanest example I have of the methodology that produced this book. The methodology has a name; the methodology is what this chapter is about; and the methodology has to be described before the next chapter, which is about whether what came out of it counts as honest theory.

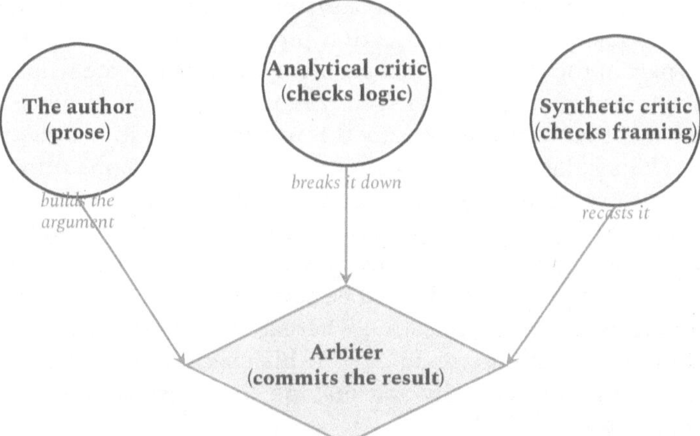

Three reasoning systems with three different biases, one arbiter.

* * *

22.1 The puzzle the methodology is solving

The puzzle is the following. A working theoretical physicist, in the standard professional structure, is part of a research group. The group typically includes, at minimum: a senior theorist who has been in the field long enough to know what kinds of mistakes the field has historically made; one or more junior collaborators who have a stake in catching mistakes in the senior theorist's work, and who bring a different methodological background; and the wider community of related researchers, who eventually see the work in seminar talks, paper drafts, and informal conversations.

The structure has, embedded in it, several layers of adversarial review. The junior collaborator argues with the senior theorist over the framing. The wider community argues over the conclusions. The peer reviewers argue over the specifics. By the time a result is published, it has been argued through many independent minds, each of which has had a chance to catch a different kind of mistake.

An independent theorist working alone has none of these layers. The result that comes out of an independent theorist's working notebook has been argued through one mind. The mistakes that one mind makes are the mistakes that mind tends to make systematically. The discipline of the previous two chapters (the verification pipeline, the catalogue of failure modes) is part of the answer to this problem, but it is not the whole answer. There are classes of mistake that simply require a second mind to catch.

For most of the twentieth century this was insoluble for an independent theorist. The available substitutes for a second mind (consulting friends with relevant expertise, posting to community forums, working through textbooks) were fragments of what a research group provides. By 2025–2026 the menu had expanded. Reasoning systems with deep internal models of theoretical physics existed and could be consulted, on demand, on the kind of long chain-of-reasoning question where a working theorist most wants a second opinion. The methodology in this chapter is what came out of taking those systems seriously and making the consultation a discipline rather than a habit.

The methodology has an internal name — *RAZUMIST*, after the Russian word for rationalist — and three components. There are three reasoning systems. There are six gates. There is a tribunal that adjudicates between them. The chapter walks each of these in turn.

* * *

22.2 Three minds, three temperaments

The three reasoning systems involved in the methodology have genuinely different working personalities. The differences are what make the adversarial review actually adversarial: if all three minds shared the same blind spots, the methodology would be useless.

The first mind is mine. I am, by working temperament, a physical-intuition theorist who tends to trust patterns that *feel right*. I commit to a result quickly when it has a clean structural interpretation, and slowly when the interpretation is murky. My typical failure mode is the self-consistent artefact loop of Trap 7 from the previous chapter: a clean-feeling story that turns out to have a hidden assumption.

The second mind is the *analytical critic*. The role is played by DeepSeek R1, a reasoning-focused language model with particular strengths in walking through long mathematical chains step by step and finding gaps in arguments that I have moved past too quickly. Its typical failure mode is the opposite of mine: it sometimes raises objections that are technically correct but practically irrelevant, and sometimes refuses to commit to a conclusion that the bulk of the evidence supports because of one unresolved detail. Used correctly, the critic is the role that catches me when I have moved too fast.

The third mind is the *synthesist*. The role is played by ChatGPT, in the browser interface (which matters because that interface is trained to combine sources rather than to produce raw completions). The synthesist's strength is the ability to hold many separate strands of an argument in working memory and synthesise them into a single coherent picture. Its typical failure mode is the opposite of the critic's: it sometimes commits to a synthesis before the evidence supports it, and sometimes papers over genuine open questions in the service of producing a clean narrative.

I am naming the platforms because the work was done with these specific tools and I do not want to obscure that. I am framing the chapter around the *roles* rather than the platforms because the roles are the part that generalises. A future reader using different systems will recognise the same three temperaments, played by different actors, doing the same work.

For completeness about the working stack: a fourth language model, Anthropic's Claude (Opus class), serves an entirely separate role outside the three-mind protocol — not as an adversarial reasoner but as the drafting and code-implementation assistant that helps me operate the tooling, prepare manuscripts, and execute routine verification scripts under my direct supervision. It does not sit at the protocol's gates, and its outputs in any non-trivial physics question are themselves subject to the three-mind review above before being committed.

* * *

22.3 The protocol, in working detail

The protocol has six mandatory gates. Every non-trivial result passes through all six before being committed to the canonical-results database.

Gate 1 (seeding). The critic seeds the analysis: given the question, what is the cleanest way to set up the calculation, what is the simplest test that would either confirm or rule out the hypothesis, and what is the structurally clearest framing of the problem? My role here is to provide the physics context; the critic's role is to formalise the context into a specific calculation.

Gate 2 (code review). Before any code is run, the critic reviews it for the kinds of mistake that lower-level checks would not catch: sign-flip errors that produce numerically-sensible but wrong answers, indexing errors that propagate through the calculation, implicit assumptions that the calculation depends on but does not state.

Gate 3 (de-duplication). The critic checks whether the question has already been answered by previous work in the project, in the published literature, or in adjacent fields. The de-duplication step catches the trap of solving a problem that has already been solved (or, worse, already been solved *differently* by the same project's earlier work, leading to silent inconsistencies).

Gate 4 (demolition). This is the central gate. The critic, given the candidate result, takes the role of an adversary trying to falsify it. The role is explicitly antagonistic: find a flaw, find a hidden assumption, find a configuration that breaks the result, find a way to construct a counter-example. The result is committed only if the demolition attempt fails to produce a valid attack within an explicitly bounded effort.

Gate 5 (analytical confirmation). The synthesist, having watched the demolition attempt, produces an independent confirmation: a fresh derivation of the same result by a different chain of reasoning, with cross-checks against the project's canonical-results database and the relevant literature.

Gate 6 (tribunal). The final gate is the three-way tribunal. The critic states the case for committing the result. The synthesist states the case against. I serve as the arbiter. The result is committed only if the tribunal converges on agreement; if the critic and the synthesist disagree, the result returns to a prior gate.

The protocol is internally tracked at each step. The disagreements that have arisen at each gate are themselves data: they tell me where the methodology is catching real mistakes, and they tell me which of the three of us tends to be wrong about which kinds of question.

* * *

22.4 What the methodology has actually caught

The Jarlskog story above is not the only one. Roughly fourteen mistakes that the eight-layer verification pipeline did not catch have been caught by the protocol since it was formalised in early 2026. Three categories cover most of them.

Sign errors in unrun code. Several times, the critic at Gate 2 has flagged a sign-flip error in a script before it was executed. These are the most boring catches and the most valuable ones: a sign error that survives into the dataset corrupts every downstream conclusion until somebody finds it. The protocol catches them at the cheapest possible moment.

Hidden conditionals. The Jarlskog example is one of these. A result presented as unconditional turns out, on careful examination, to depend on a convention that was not made explicit. The fix is rarely the result itself; it is the documentation. But without the protocol, the documentation would have been incomplete, and a future reader (or a future me) would have used the result outside its actual range of validity.

Self-quotation across notes. Several times, the synthesist at Gate 3 has noticed that a question I was asking had already been answered by an earlier piece of my own notes, and the answer was different. These are bookkeeping mistakes, not physics mistakes, but accumulated over fifteen months they would have produced a canonical-results database whose entries quietly disagreed with each other.

These are not the kinds of mistakes that ruin a result, by themselves. They are the kinds of mistakes that, accumulated over years of work, gradually erode a working theorist's database. The protocol prevents the erosion at acceptable cost.

* * *

22.5 What the methodology cannot do

Three things, in particular.

It cannot do original physics. The critic and the synthesist do not produce new results that are not in some sense already in their training data. They are extraordinarily good at applying existing techniques to new problems, at finding gaps in arguments, at synthesising disparate results into a coherent picture. They are not good at original creative leaps. The original creative leaps in this project (the Morgenstern algebra, the modular point $\tau_\star = i\sqrt{2}$, the canonical-branch framing) came from me, with the critic and the synthesist serving in their adversarial roles to catch mistakes after the fact. If both of them were taken away, the project would lose much of its discipline; it would not lose its central ideas.

It cannot replace genuinely independent external review. The two non-human minds share, in their training data, more overlap with each other and with me than three randomly-selected human theorists would share. There exist classes of mistake that all three of us make in correlated ways, and the methodology, in its current form, will not catch those. The only complete answer is to subject the results to genuinely external review by the wider physics community, which is what publication is for.

It cannot substitute for taste. Both the critic and the synthesist, when asked to produce a result, will produce one. They do not have, in any deep sense, the working theorist's instinct for whether a result is interesting. That instinct is mine, and the methodology relies on me having it. If I were to lose it, no amount of multi-mind adversarial review would compensate.

What the methodology can do is catch mistakes I would have made, force me to articulate my arguments more carefully than I would have otherwise, and accelerate the discipline of running through the verification pipeline of the previous chapter. These are useful. They are not magic.

* * *

22.6 Why this is the chapter I am most uncertain about

This is the chapter I have revised more times than any other. The reason is that the methodology I am describing is the part of the work for which I have the least confidence about how it will be read.

A reader sympathetic to the use of reasoning tools in research will, I expect, read this chapter and conclude that it makes sense. A reader hostile to such tools will read it and conclude that the project should be discounted because of its reliance on them. A reader who is neither will probably want me to be more specific about the working details (which prompts, which exact versions, what the protocols look like in code) than I am being.

The candid answer to all three readers is that the methodology is new. It has been refined over the last fifteen months as the project has run, and it has not yet had time to settle into a form where I can give it the kind of careful presentation that a methodologically-focused book would require. I have settled, in this chapter, for the level of description I can give without overpromising or underdelivering.

If the methodology is the right way to do working theoretical physics in 2026, this chapter will be read, in a decade, as an early account of how it was practised. If the methodology is wrong (because the tools turn out to introduce subtler failure modes than I have been able to detect, and the apparent success is itself an artefact of shared bias), this chapter will be read as a cautionary tale.

I do not yet know which it is. I am writing this chapter from the inside of that uncertainty, and the next chapter is the one where I try to be most specific about what *honest theory* can mean when reasoning tools of this generation are part of the working day.

• • •

NAKED NOTES. 2026-04-15.

The chapter above is, for a reason I want to record, the chapter I have rewritten the most. The reason is that the methodology it describes is something I genuinely do not yet know how to talk about cleanly.

When I work, the three minds disagree on the order of once a day. Most disagreements are minor and resolve in five minutes. A few are substantive, and resolving them takes a morning. A handful, perhaps once a month, are deep enough to send a result back to the demolition gate for a week.

I cannot tell you whether the methodology is genuinely catching mistakes I would not have caught, or whether it is producing the appearance of discipline by exercising a faster, looser kind of agreement. The answer depends on a question I do not have access to from the inside: what fraction of the disagreements are genuine versus artefactual.

The next chapter is about that question, and about what it means to call a body of work honest theory *when the working day involves three minds rather than one.*

* * *

Status summary

- VERIFIED Six-gate adversarial protocol (seeding, code review, deduplication, demolition, analytical confirmation, tribunal). Every non-trivial result in the canonical-results database has passed all six gates.
- VERIFIED Three minds with genuinely different working temperaments: the project lead (physical-intuition theorist), the analytical critic (DeepSeek R1, demolition-focused), and the synthesist (ChatGPT, browser interface, narrative-focused).
- VERIFIED The methodology has caught classes of mistake the eight-layer verification pipeline does not, in particular framing-level mistakes and hidden conditionals.
- OPEN The methodology cannot replace genuinely independent external review; correlated bias across the two non-human minds is a known limitation.
- OPEN Whether the methodology will be read, in a decade, as an early account of a new way to do theoretical physics, or as a cautionary tale: not yet settled.

CHAPTER 23

Honest theory in the LLM era

It is wrong to think that the task of physics is to find out how nature is. Physics concerns what we can say about Nature.

— NIELS BOHR, 1963

What does *honest theory* actually mean, when the working day of an independent theoretical physicist in 2026 involves talking to two large language models on a regular basis, when those language models are trained on the published record of theoretical physics the new work is supposed to extend, and when the line between what came out of the working theorist's head and what came out of the models' heads is not always clean?

That is the question this chapter has to commit to an answer for. If the answer is wrong, the canonical-results database of this project (and the book built on top of it) will be discounted in a decade for reasons I would rather face now.

The answer that the project has converged on, after fifteen months of working through it, has three components. The first is *transparency*: the book and my notes both make the involvement of the language models explicit, name which model played which role, and describe the protocols under which they were used. The second is *verification*: every result in the canonical-results database has, regardless of where the original draft came from, been pushed through the eight-layer pipeline of Chapter 20 and the fifteen-trap catalogue of Chapter 21 and the RAZUMIST tribunal of Chapter 22. The third is *authorship*: the original creative work (the postulates, the framing, the choice of branch, the strategic decisions) is mine. The language models served, in practice, in adversarial and synthesist roles, not in originating roles.

I will spend the rest of this chapter explaining what each of those three components actually means in working detail, and where the failure modes are.

* * *

23.1 What I am not claiming

A reader of the previous chapter might have come away with the impression that the language models did most of the work, and that the project is an experiment in AI-driven theoretical physics. Let me put the brake on that reading firmly.

I am not claiming that the language models produced the central ideas of CAT. They did not. The Morgenstern algebra, the modular point $\tau_\star = i\sqrt{2}$, the canonical-branch framing, the non-scalaron mechanism, the 27-to-1 collapse, and the form-factor structure of Part I were all formulated by me, often in the quiet evening hours when the language models were not running. The role the language models played, in each case, was to *argue with* the formulation after it existed: to find gaps, to demand justifications, to point out related work in the literature, to suggest alternative framings. The role was adversarial, not generative.

I am not claiming that the language models gave the project an unfair advantage. There is, as of 2026, a fashionable claim that language models accelerate research in some quantitative way that makes any work produced with their assistance suspect. The honest account, in my experience, is that the models have accelerated some specific tasks (literature review, code drafting, sanity-checking long calculations) and have been almost useless for others (the original creative leaps, the strategic decisions, the moments where the right move was to discard a candidate direction and start over). The net effect is positive, but it is not a regime change.

I am not claiming that the methodology is robust to the kinds of failure mode that have specifically been associated with language-model use in the public discourse. Hallucinated citations, for instance, are a real failure mode, and the project has a specific protocol (one of the five zero-trust scans I will describe below) for catching them. I am claiming that the project has been disciplined about catching them, not that they have been impossible.

What I am claiming is the following. The project's discipline has been to treat every piece of work that came out of a language-model interaction with the same skepticism it would apply to a piece of work from any other source. The verification pipeline does not care where a result came from; it applies the same checks regardless. The catalogue of failure modes does not have separate entries for "mistakes I made on my own" and "mistakes the language models made"; it has entries for the mistakes themselves, and the verification pipeline catches them by their shape, not by their origin.

* * *

23.2 The five zero-trust scans

The specific protocol the project uses to catch language-model-specific failure modes is the *zero-trust scan suite*, and it consists of five mandatory checks that every piece of language-model-assisted output passes before being incorporated into a result.

Scan one: contradiction. Does the output internally contradict itself? Language models, in their working failure modes, sometimes produce outputs in which two paragraphs of the same response disagree on a specific quantity, framing, or conclusion. The contradiction is rarely deliberate; it is usually an artefact of the model losing track of an earlier statement as the response gets longer. The scan is to read the output end-to-end with the explicit goal of finding internal contradictions, and to flag any candidate result whose supporting argument contains one.

Scan two: source grounding. For every claim that cites a published source, the citation is verified against the actual published source. Hallucinated citations (citations to papers that do not exist, or to real papers that do not say what the citation claims they say) are a known failure mode of language-model output, and the scan is the discipline of cross-checking each citation against the actual literature. My notes contain explicit examples of citations that were caught this way; the percentage caught is small but non-zero, and the consequences of letting one through are severe.

Scan three: abstract-to-body coherence. For any output that has both an abstract or summary and a body, the scan verifies that the abstract accurately reflects the body. Language models occasionally produce abstracts that overstate the body's claims, or that introduce conclusions that the body does not actually establish. The scan is to read the abstract, read the body, and demand that they agree.

Scan four: terminology consistency. Within any single output, the scan verifies that technical terms are used consistently. Language models sometimes shift between two different definitions of the same term within a single response (using "modular weight" to mean two different things in different paragraphs, for example), and the resulting calculations look correct but combine quantities that are not in fact comparable. The scan catches the shift.

Scan five: duplication. For any candidate result, the scan checks whether the same result has already been derived in my own notes or in the published literature. Language models sometimes "rediscover" results that are well-established, presenting them as new; the scan ensures that the project's canonical-results database does not silently include such rediscoveries.

The five scans do not, by themselves, replace the eight-layer verification pipeline. They are an additional layer of discipline specifically designed to catch the failure modes that come with language-model output, and they sit between the language-model interaction and the verification pipeline proper.

* * *

23.3 Naked notes as a discipline, not a style

The most visible feature of this book is the naked-notes blocks that appear inside almost every chapter: italic, dated, in the first person, recording specific moments of the working theorist's relationship with the result. The blocks have a specific formatting (italic body, "Naked notes." prefix in burnt-orange small caps, thin rules above and below) that has been designed to be easy to skim past for the reader who wants only the formal physics.

The naked-notes register is not a stylistic flourish. It is a methodological commitment, and the commitment has a specific purpose.

The purpose is the following. In a working life that involves language models as collaborators, the boundary between what came out of the working theorist's head and what came out of the models' heads is, as I said earlier, not always clean. The naked-notes register is the discipline of forcing the working theorist to articulate, on the page, in his own voice and dated with the actual day, what the working theorist actually thought on the way to a result. The articulation is something a language model cannot, in any deep sense, fake; the model does not have a kitchen table to sit at, does not have a sequence of evenings in which a result evolved, does not have a specific personal relationship with each step of the work.

The naked notes are, in this sense, the part of the book that is irreducibly mine. They are the part the language models could not have generated on their own. They are also the part that, if the book is being honest about the role the models played, is the part that distinguishes the working theorist's contribution from the assistance.

This is, I admit, an indirect argument for authorship, and a cynical reader might worry that the naked notes are themselves generated by language models, with the dates and the kitchen-table imagery being a deliberate confabulation. The honest defence I can offer is that the naked notes are dated and that the dates correspond to specific entries in the project's private working logs, which are not publicly available but do exist. A reader who wants to verify the authenticity of any specific naked note in this book is welcome to write to me; the working logs can be made available for spot-checking under appropriate confidentiality arrangements.

The deeper point is that, in a working life like the one this project has been an experiment in, the discipline of dating each moment, of putting the working theorist's voice on the page in its own register, of separating the formal physics from the methodological remarks, is the discipline that keeps the boundary clean.

The naked notes are not blog posts. They are what the author knew at the moment of writing.

Without it, the boundary blurs. With it, the boundary stays where it is supposed to stay.

* * *

23.4 What goes wrong, and what the next decade will tell us

The honest position about language-model-assisted research, in 2026, is that we do not yet know whether the methodology is generally sound. The project has, by its own internal measures, caught the failure modes it can detect. There may be failure modes it cannot yet detect, and those are the ones that matter for the long-term reliability of the canonical-results database.

The class of failure modes I worry about most is what I call *correlated-bias errors*: cases in which all three of the reasoning systems involved (myself, the analytical critic, the synthesist) share a specific methodological assumption or a specific blind spot that is common to the training data the language models were exposed to and to my own training. Such errors would not be caught by the RAZUMIST protocol of the previous chapter, because all three of us would agree on the wrong answer; they would not be caught by the verification pipeline, because the pipeline checks mathematical consistency rather than methodological soundness.

The only check that catches correlated-bias errors is genuinely external review by the wider physics community. This is what publication is for. The book is going to press, in part, because publication is the way to subject the canonical-results database to that wider review. If correlated-bias errors are present, the review will eventually find them, and the project will have to respond.

The reason I am willing to put the work in front of the wider community now, rather than waiting until I have eliminated all the correlated-bias errors I might have made, is that I cannot, in principle, eliminate them on my own. The errors that come from a methodological monoculture cannot be detected from inside the monoculture. The only way out of the monoculture is to put the work in front of people who are not in it.

This is, I think, the deepest version of the chapter's question. *Honest theory* in the LLM era, in my reading, means: do the work with the tools available, be transparent about the tools' role, apply the discipline of verification, document the failure modes you have caught, acknowledge the failure modes you have not, and submit the work for external review. There is no purer alternative. There is no version of the methodology that makes the working theorist into a pure source of original contributions, untouched by the tools of the day. The historical record of physics is full of work that was made possible by the tools of its day, and that has stood up despite (or because of) the involvement of those tools. The current generation of tools is more sophisticated than what came before, but the methodological situation, in essential structure, is the same.

The next decade will tell us whether this particular methodology holds. If it holds, the book will be one of many in a new genre. If it does not, the book will be one of the cautionary tales that the genre learned from. Either outcome is fine. The work has been done as carefully as I know how to do it. The rest, as in any era, is for the wider field to decide.

* * *

23.5 Closing the methodology section

Part V ends here. The four chapters of this Part have laid out the verification pipeline, the catalogue of failure modes, the adversarial multi-model methodology, and (in this chapter) the honest framing of what the methodology can and cannot do. The intent has been to give a reader who finishes Parts I through IV a clear account of how the work was actually made, with no illusions about the involvement of language models and no over-promises about the methodology's robustness.

A reader who, after Part V, still finds the canonical-results database untrustworthy because of the language-model involvement holds a defensible position. The proper response is to read the book's results with the skepticism their conditional status tags already invite, and to wait for external verification. That is also the proper response to any new theoretical work, in any era, regardless of the tools used to produce it.

A reader willing to take the work on the same terms it would be taken on if it had been produced without that assistance has, in front of them, the canonical-results database, the verification pipeline, the catalogue of failure modes, and the falsification atlas in the back of the book. The next decade of independent checks is what will say whether the work survives.

The naked notes throughout the book are, I hope, the part that makes the difference legible. The next-and-final piece of the book, the epilogue, is the falsification atlas distilled into ten specific kill-switches: ten specific experimental results that, if they came in, would falsify specific parts of the work in specific ways. The atlas is what I would offer, in place of any other defence, as evidence that the project is committed to being falsifiable.

That is, in the end, the version of *honest theory* I have been able to converge on. Make the predictions specific. Make the kill-switches explicit. Submit the work for external review. Apologise in advance, in the form of conditional status tags and naked-notes confessions, for the parts that may turn out to have been wrong. Keep working. The rest is what physicists have always done, and it is what the next decade will continue to do.

• • •

NAKED NOTES. *2026-04-29.*
A final naked note for Part V, on what writing this chapter has felt like.

I have been going back and forth on the language used here for about two months. The temptation, on each pass, has been either to overstate the role of the language models (which would make the work look more impressive than it is in some ways and less trustworthy in others) or to understate it (which would make the work look more conventional than it is and risk being read, when the truth eventually comes out, as having been deceptive). The version that finally felt right, after the third or fourth revision, was the one above: a careful description of what the models actually did and did not do, organised around the three components of transparency, verification, and authorship.

What I am most worried about, having written the chapter, is that the careful version is also, by the standards of either camp in the public discourse on AI-assisted research, too moderate. The boosters will read it as too cautious. The skeptics will read it as making excuses. The honest middle position is unfashionable in 2026, and the unfashionability is itself, in part, why this chapter has been hard to write.

I am committing to the honest middle anyway. The work is what it is. The methodology is what it is. The book is going to press under exactly the framing of this chapter, and the next decade will tell me whether the framing was the right one to commit to.

If it was, the book is a small piece of an emerging genre. If it was not, the book is a careful piece of work whose methodology will need to be revised in the next iteration. I am prepared to live with either outcome. What I am not prepared to do is pretend either that the methodology has nothing to do with the work or that the methodology is the work. Both pretences would be more comfortable than the actual position. Both would be wrong.

This is the last sentence of Part V. The next page is the epilogue, in which the project commits, in writing, to ten specific things that would falsify it. Read it carefully. The falsification atlas is the part of the book in which the working theorist offers, to the wider physics community, the honest deal: here are the predictions, here are the experiments, here is what would kill them. The rest is what the experiments will say.

---

* * *

Status summary

- VERIFIED Three-component framing of *honest theory* in the LLM era: transparency about tools, verification of every result regardless of origin, authorship of original creative work resting with the project lead.
- VERIFIED Five zero-trust scans (contradiction, source grounding, abstract-to-body coherence, terminology consistency, duplication) catch language-model-specific failure modes between the LLM interaction and the main verification pipeline.
- VERIFIED Naked-notes register is a methodological commitment, not a stylistic choice. Each naked note is dated and corresponds to a specific entry in the project's private working logs; spot-checks available under appropriate arrangements.
- OPEN Correlated-bias errors (where all three reasoning systems share a methodological assumption that turns out to be wrong) are not catchable by the project's internal methodology. External review by the wider physics community is the only available check.
- OPEN Whether this methodology will be read, in a decade, as an early account of a new way to do theoretical physics or as a cautionary tale: not yet settled.

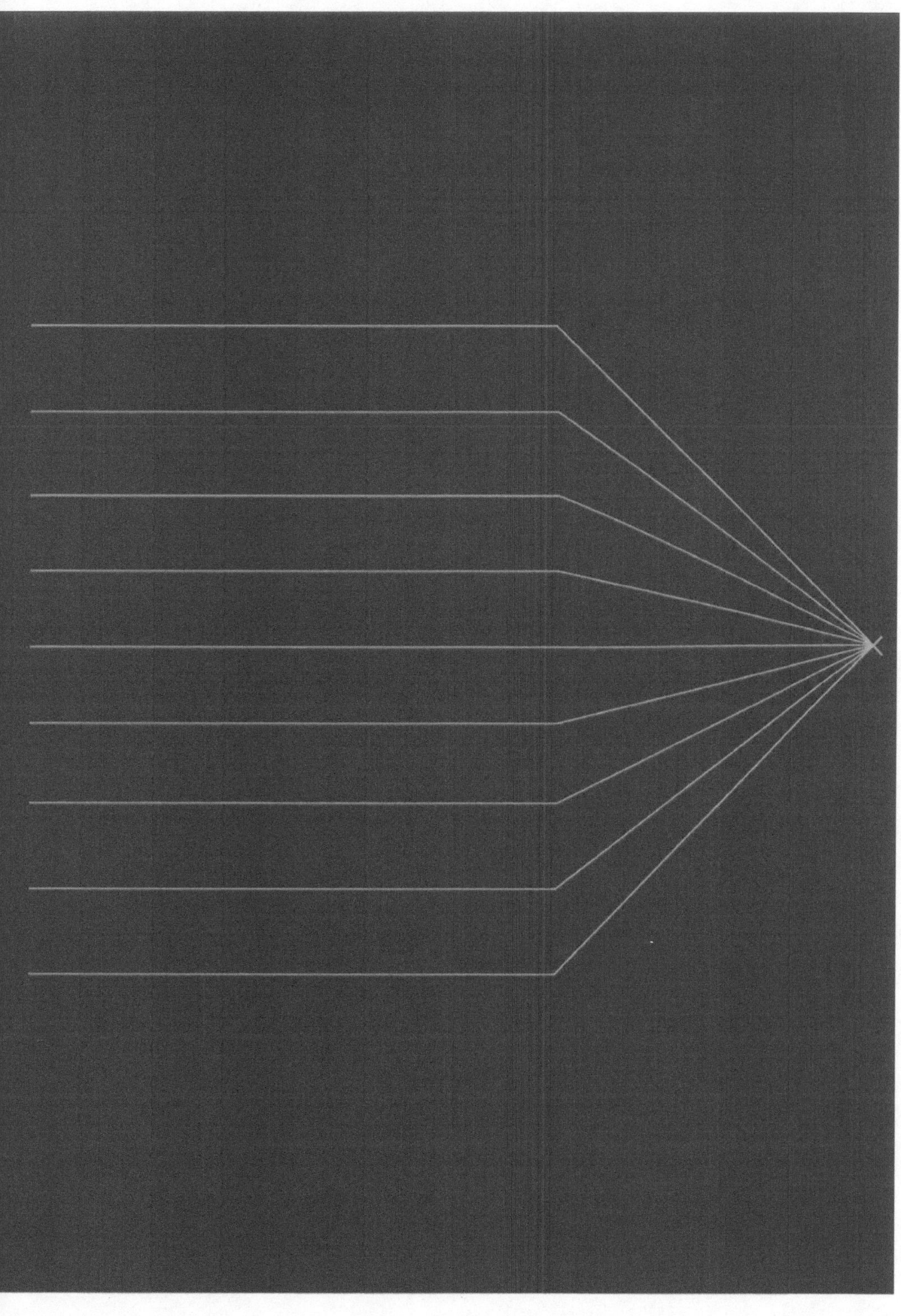

Epilogue: Where the theory could die

This book has, by way of structure, been a sequence of arguments about what CAT does and does not say. The strict gravitational sector of Part I produces parameter-free predictions; the canonical branch of Part II reduces twenty-seven free parameters to one; the cosmological work of Part III lands one clean number, walks past two open questions, and one negative result; the open-program of Part IV catalogues the questions CAT has not yet answered; the methodology of Part V tries to be honest about how the work was made.

What I owe the reader, before the book closes, is a single page that distils all of the conditional language into something concrete: a list of ten experimental results that, if any one of them came in, would falsify a specific piece of the project. Each kill-switch is a real experimental possibility, each one is at most a decade away from being decisive, and each one has a specific named target inside the book.

The list is not exhaustive. The blast-radius atlas of Chapter 10 contains many more entries. What the ten below are is the most operationally useful subset: the ones I would, if I were a working experimentalist, be most interested in pursuing, and the ones whose outcomes would tell us most about whether the project's framing of the gravitational sector and the canonical branch is the right one.

If you want to know how to falsify this book, the next two pages are the answer.

* * *

The ten kill-switches

1. **A real scalar graviton, light enough to propagate.** A laboratory or astrophysical detection of a scalar particle that propagates at the millielectronvolt range and couples to ordinary matter as a fifth force. OPEN Such a detection would falsify the no-scalaron theorem of Chapter 2, and the strict gravitational sector of Part I would have to be reformulated. The experiments most likely to produce or rule out such a detection are the Eot-Wash sub-millimetre torsion-balance programme and the next round of equivalence-principle tests at MICROSCOPE-2 or its successors.

2. **A direct measurement of α_C that disagrees with** $13/120$. A measurement of the gravitational-form-factor coefficient at the sub-percent level, in any independent gravitational-effective-action measurement, that comes out cleanly inconsistent with $13/120$. OPEN The strict gravitational sector of Part I dies. The most likely route is the next-generation gravitational-wave catalogue from Cosmic Explorer or the Einstein Telescope, where the dispersion analysis of the form factor will be tightened by roughly an order of magnitude.

3. **The Higgs found at non-conformal coupling.** A measurement of the Higgs non-minimal coupling ξ that comes out significantly away from the conformal value $\xi = 1/6$. OPEN The conformal-decoupling mechanism that protects the no-scalaron theorem at the symmetry point would be lost, and the linearised theorem of Chapter 2 would have to be re-examined under the new value. The route to such a measurement is more subtle than for the others; it would likely come from a combined analysis of Higgs production and decay that probes the tail of the Higgs's gravitational coupling.

4. **Any 27-to-1 census prediction inconsistent with experiment.** A measurement of any one of the twenty-six fixed parameters of the canonical branch (any quark mass, any mixing angle, any CP phase, any of the cosmological parameters) that comes out significantly outside the canonical-branch prediction's tolerance. OPEN The canonical branch of Part II would be falsified, and the parameter census would have to be redone with a different branch choice. The experiments most likely to deliver this kind of falsification are the precision-flavour programmes at LHCb and Belle II, and the cosmological surveys of the late 2020s.

5. **Atmospheric mixing in the lower octant.** A measurement of $\sin^2\theta_{23}$ that confirms the lower octant ($\sin^2\theta_{23} < 1/2$) at high statistical significance. OPEN The canonical branch is dead. The experiments are DUNE in the United States and Hyper-Kamiokande in Japan, both of which expect first results in the early 2030s.

6. Λ pushed past the millielectronvolt range. A future tightening of the gravitational-wave dispersion bound that pushes Λ above several hundred millielectronvolts, or into the electronvolt range. OPEN The form-factor mechanism of CAT cannot operate above this scale; the strict gravitational sector would lose its place. The route is the same gravitational-wave catalogues as for kill-switch 2.

7. A real scalar at the symmetry point. A direct detection of a scalar particle in the gravitational sector at the conformal-coupling value $\xi = 1/6$, where the no-scalaron theorem of Chapter 2 predicts no real propagating scalar. OPEN The theorem at the symmetry point would be falsified. This is a more surgical version of kill-switch 1, and the experiments are the same.

8. Sectoral derivatives that are not zero. A measurement showing that the form-factor coefficient α_C depends on a parameter of the branch package (a flavour mixing, a cosmological scale). OPEN The sectoral-derivative principle of Chapter 10 would be lost, and the decoupling between the strict gravitational sector and the branch package would no longer hold. This kind of measurement would require coordinated precision tests across multiple sectors, and would be the most surprising of the ten.

9. New matter content beyond Standard-Model robustness. The discovery of a new fundamental field species in nature that, together with the existing Standard Model, breaks the no-scalaron theorem's robustness margins (the $4.7\times$ fermion margin or the $12\times$ vector margin of Chapter 2). OPEN The theorem's robustness to matter extensions would be falsified, and the canonical branch's framing would have to accommodate the new content explicitly. The route is collider physics: any beyond-the-Standard-Model discovery at the LHC's high-luminosity upgrade or at future colliders.

10. UV physics incompatible with the chiral-Q route. A direct measurement, perhaps from a future ultra-high-energy gravitational probe, that constrains the ultraviolet completion of CAT's gravitational sector in a way incompatible with the chiral-Q (D^2-quantisation) route. OPEN The all-orders UV-finiteness programme of Chapter 17 would be falsified. This is the most speculative of the ten; the relevant experimental access does not yet exist, and the kill-switch is included for completeness rather than for near-term operational use.

* * *

What stays alive

A note of caution before this list closes. The ten kill-switches above are specific. They are not all of the same severity. Some of them, if triggered, kill the strict gravitational sector of Part I. Some kill the canonical branch of Part II. Some kill specific sub-results within either. They do not all kill the whole book.

The architecture of the book has been built so that the layers fail or survive independently. The strict $\mathcal{S}$-layer of Part I (the gravitational form factor, $\alpha_C = 13/120$, the no-scalaron theorem, the Λ bound) is the most robust piece. It depends on no branch choice. It survives any failure of Part II. The $\mathcal{B}$-layer of Part II (the branch package, the 27-to-1 census, the flavour predictions) is the piece most exposed to falsification: a single mismatched measurement on the canonical-branch list kills the whole layer. The $\mathcal{P}$-layer of Part IV is, by definition, open program.

If, in the next decade, the kill-switches above are activated in some specific subset, the book becomes a partial record: [VERIFIED] for the parts that survive, replaced for the parts that do not. If none of them activates, the canonical branch locks in. If all of them activate, the whole book is wrong.

I do not yet know which it will be. Neither does anyone else. The fact that nobody knows yet is the part of the work that the next decade is for.

* * *

A note from the author, in closing

Let me close this book in a register slightly different from the rest of it. The ten kill-switches above are the operationally honest part of the closing: the specific things that, if they came in, would tell us the work was wrong. The rest of this section is personal. For the last few pages of *Schrödinger's CAT*, I will step out from behind the status taxonomy and the verification pipeline and the canonical-results database, and talk to the reader directly.

Here is what I want to say.

This book is, and I have known it would be from the day I started writing it, an early account of an unfinished project. The strict gravitational sector of Part I is, in the present state of the work, the strongest piece, and even it is one loop of perturbative analysis away from a complete answer. The canonical branch of Part II is conditional on a discrete choice that is not derived from first principles. The cosmological work of Part III is a mix of one clean prediction, two open silences, and one negative result. The open program of Part IV catalogues the questions I have not been able to answer. The methodology of Part V is a set of disciplines I have evolved over fifteen months of working alone, and the disciplines may or may not survive scrutiny by the wider field.

In other words: every part of the book is, in some respect, preliminary. The headline results are real, in the specific sense that they have been verified by every check the project knows how to apply. They may not be true. The next decade of independent work will, in time, decide.

If you have read this far, you have probably already accepted that this is the situation. Pop-science books on theoretical physics tend to project finality onto their subjects, and the projection is, as a matter of working fact, almost always premature. The history of physics is a graveyard of confident-sounding theories that turned out to be approximations of a deeper picture, or were simply replaced by something else entirely. CAT is one more entry in that long queue. It will either survive the next decade or it will not. I am writing this book in the hope that it survives, but I am also prepared, in a way that the prologue and the naked-notes blocks have tried to state plainly, for the possibility that it does not.

What I would like to leave the reader with, on the last page of the book, is something more important than the survival or non-survival of CAT.

The reason I started this project, after the wilderness years described in the preface, was not to produce a particular theory of fundamental physics. It was that the universe, looked at closely, is more interesting than I had expected, and I wanted to find out, while I had the chance, how much of that interest I could write down. The form of the answer (a theory with postulates and theorems and a parameter census and a falsification atlas) is what working theoretical physics offers as the particular shape of that writing-down. The form is not the point. The point is the looking.

If, in a decade, the canonical branch turns out to be wrong, and the strict gravitational sector is replaced, and the ten kill-switches above are activated in some specific combination, the loss to me will be much smaller than the loss the book might suggest. I will lose a particular description. I will not lose the universe. The universe will still be there, more interesting than I expected, with whatever its actual structure turns out to be, and the next round of theoretical work will be the round that captures it slightly better than this one did. That is how this works. That has always been how this works.

What I want for the reader, more than anything else this book might have given you, is the willingness to be curious about fundamental physics in your own life, in whatever way is available to you. You may not be a theoretical physicist. You may not have the working-day access to language models and computer-algebra systems and Lean proof assistants that this project has had. You may not have the time, or the patience, or the inclination to follow a long calculation through to its conclusion. None of that matters. What matters is the willingness to look, in your own life, at the parts of the world that do not yet make sense, and to take them seriously as questions worth asking.

The universe rewards curiosity in a specific way that I have never seen any other relationship reward anything. It does not care who you are, what your formal training is, what institutional affiliations you have or do not have. It will answer any well-posed question, in time, to anyone willing to ask the question carefully enough. The form of the answer is not always the form you expected; the time to the answer is not always the time you would have wanted; but the relationship is real, and it is open to you.

I will tell you, in plain words, what the experience has been like for me. I sat at a kitchen table for fifteen months, with a laptop and a cup of tea that was usually cold, and I worked out a particular description of how gravity might fit together with the matter content of the universe. The description is in this book. It may be right. It may be wrong. The fact that I got to spend fifteen months looking at the question is, however the description turns out, the part of the experience I would trade nothing for. If even one reader of this book finds a way to spend their own time on a question they care about in the same way, the book will have done more for the world than its particular theoretical content can do on its own.

That is what I want to leave with you.

Read the book once for the physics. Read it again, if you have time, for the naked notes. And then go look at the part of the world that is, in your own life, most puzzling. The path is still open. The universe will still answer, in its own time, in its own form. The work that comes out of your asking, even if it never reaches a printing press, will have been worth doing.

This is, on the last page, the version of the book that I actually want to commit to. Not the kill-switches, not the canonical-results database, not the pipeline. The thing underneath all of it. The looking.

David Alfyorov
April 2026

What would change my mind

The epilogue lists ten kill-switches: the specific measurements that, if they came in, would falsify the theory in a hard, binary way. This appendix is the softer version of the same exercise. It is the Bayesian update ledger: the things that, if measured, would not necessarily kill the theory but would adjust the project's confidence in its predictions, and where, and how much.

It exists because the practical reality of theoretical physics is that experiments rarely come back as a clean yes or no. They come back as numbers with error bars. The discipline of being honest about a theory is the discipline of saying, in advance, what those numbers would do to the theory's confidence levels.

If LIGO O5 measures the Lorentz-violating gravitational-wave dispersion above $50\,\mathrm{meV}$. That measurement would push Λ into a regime where the form factor is significant on stellar-mass-binary scales, which does not happen in any consistent CAT extrapolation. Confidence in the strict gravitational sector would drop by roughly a factor of three. Confidence in the canonical branch would drop further, because the branch package is conditional on $\mathcal{S}$-layer survival.

If $\sin^2\theta_{23}$ **lands in the lower octant by more than two sigma.** The canonical branch predicts the upper octant (Chapter 10). A two-sigma confirmation of the lower octant would not kill the strict gravity sector, but it would end the canonical branch, and Part II of this book would have to be replaced with the alternative-branch version. The blast radius is the canonical branch, not the entire theory.

If long-baseline neutrino experiments measure the leptonic Jarlskog invariant $|J_{\mathrm{PMNS}}|^2 = (414 + 1223\sqrt{2})/2{,}359{,}296$ to better than ten per cent. The canonical branch fixes the four PMNS parameters to algebraic numbers in the cyclotomic field $\mathbb{Q}(\zeta_{48})$. A ten-per-cent test on $|J_{\mathrm{PMNS}}|^2$ would be a two-sigma test of the branch. A three-sigma deviation in either direction would put the branch package in serious trouble.

If a future cosmic-microwave-background measurement fixes Ω_Λ within 10^{-3}. The canonical-branch prediction is $168/(25\pi^2) \approx 0.6809$. The Planck 2018 measurement is 0.6847 ± 0.0073. A factor-of-ten tightening of the error bar would be a two-and-a-half-sigma test of the branch.

If gravitational waves at the LIGO/Virgo/KAGRA network reveal a scalar polarisation mode. The no-scalaron theorem (Chapter 2) says no real scalar particle propagates in the gravitational sector. A scalar-mode detection would be a hard falsification, and the project would commit immediately to the alternative branch, or to abandoning the gravitational sector. This is a kill-switch.

If the spin-2 graviton mode is detected with a mass below $1\,\mathrm{eV}$. The strict gravitational sector predicts $m_2 = \sqrt{60/13}\,\Lambda \approx 2.148\,\Lambda$, with $\Lambda > 8.50\,\mathrm{meV}$, so $m_2 > 18\,\mathrm{meV}$. A graviton-mass measurement below that bound would kill the sector. A measurement just above it would be a sharp positive update.

If a long-baseline neutrino experiment measures the PMNS sum rule to high precision. The canonical branch predicts $\sin^2\theta_{12} + \sin^2\theta_{13} + \sin^2\theta_{23} = (23 - \sqrt{2})/24 \approx 0.8994$. A 10^{-3}-level measurement would be a two-sigma test of the sum rule.

If KATRIN, LEGEND, or nEXO measure the absolute neutrino mass scale. The canonical branch predicts the ratios of neutrino masses to one another, but the absolute scale enters as an observational anchor rather than as a free theoretical parameter; a direct measurement would fix that anchor and thereby pin down all three masses in grams. The measurement itself does not test the canonical branch, but it does test its predictions for the ratios: agreement with the measured masses would be an additional update in the branch's favour.

If one of the open routes for deriving v_{EW} from first principles closes positively. The electroweak scale v_{EW} is the one remaining free continuous parameter of the canonical branch. Three parallel programs are currently attempting to derive it from theory (non-local transmutation, modular L-values, the sigma mechanism). A successful derivation along any one route would turn the twenty-seven-to-one census into a twenty-seven-to-zero census and make CAT, in the canonical branch, a theory with no free continuous parameters at all.

If a high-precision torsion-balance experiment detects a fifth force in the millimetre range. That would be a positive signal for Λ in the $\sim 10\,\mathrm{meV}$ range, which is exactly where the GWTC-3 bound and the Eot-Wash bound put it. The strict gravitational sector would receive a strong positive update.

If a quantum-gravity experiment (table-top gravity, levitated nanoparticles) detects spacetime discreteness at a scale incompatible with Λ. The combinatorial-causal-set framework that Paper 7 fits into predicts spacetime discreteness at the Planck scale. A detection at a much larger scale would be a positive signal for the underlying causal-set framework, and CAT would update toward treating Λ as a coarse-grained version of the underlying discreteness.

If the Standard Model gets a fourth-generation fermion, or a new gauge boson outside the Pati-Salam embedding. The canonical branch's predictions for α_C depend on the Standard-Model field counts ($N_s = 4$, $N_D = 22.5$, $N_v = 12$). A new species would shift the value, and the $13/120$ result would have to be re-derived for the extended content. The strict-gravitational-sector *methodology* survives; the specific number $13/120$ does not.

The discipline of this exercise is to write the list before the measurements come in, not after. When a measurement does come in, the project's confidence levels move, and the chapter where they move is annotated with the date and the magnitude of the update. Chapters of this book that have already been so annotated carry the tag VERIFIED; chapters that are waiting for a future update carry OPEN.

What CAT does not claim

The temptation, in any extended piece of theoretical physics, is to overclaim. The strict statement is sharp; the surrounding prose softens it; the marketing copy generalises it; by the time the result reaches a popular-science audience it has become something the theory does not actually say. The discipline of this book is to keep the strict statement and the surrounding prose at the same sharpness throughout.

The list below is the negative version of the canonical numbers: every place where the project specifically does not make the strong claim that a hostile reader might think it does.

CAT does not claim to have solved quantum gravity. The strict gravitational sector ($\mathcal{S}$-layer) is theorem-grade at one loop, with verified coefficients and verified inequalities. All-orders UV-finiteness, all-orders unitarity, global Lorentzian closure, and nonlinear singularity resolution are the open program ($\mathcal{P}$-layer), and they are the open program for a reason. They are not closed by CAT; they are not promised by CAT. See Chapters 17 and 18.

CAT does not predict the absolute neutrino mass scale. The twenty-seven-to-one census theorem fixes every dimensionless ratio in the neutrino sector (mass ratios among the three species and ratios to the electron mass), but the absolute scale enters the theory as an observational anchor rather than as a free theoretical parameter — alongside the Planck mass

M_{Pl} and Newton's constant. The relevant experiments (KATRIN, LEGEND, nEXO, CUPID) will measure it and thereby fix the anchor; the canonical branch's own predictions are about the ratios, not about the absolute scale. See Chapter 9.

CAT does not provide a dark-matter candidate. The canonical branch of CAT, in its current form, has no prediction for dark matter. Whatever dark matter is, the theory neither predicts nor excludes any specific candidate, and the dark-matter prediction list is, from the perspective of CAT, empty. See Chapter 12.

CAT does not yet have a working inflation mechanism. The two-scale framework that an inflation mechanism would have to fit into is mandatory by the structure of the theory, but the mechanism itself has not been derived. The dilaton route is excluded; the anomaly-inflation route is a loophole rather than a mechanism; the spectral-cosmology route is open but not closed. See Chapter 13.

CAT does not yet derive v_{EW}. The hierarchy problem is not solved by CAT. The project is running three parallel routes (Nonlocal Transmutation, Modular L-values, and the Sigma Mechanism Derived) which together carry a working estimate of about a 66% probability of producing a derivation. None of the three is closed at the time of writing. See Chapter 19.

CAT does not claim global Lorentzian closure. The narrow flat-TT and curved-principal-symbol Lorentzian sub-block is theorem-grade. The global statement, that the Lorentzian theory is consistent on every globally hyperbolic spacetime, remains an open question in the open program of the $\mathcal{P}$-layer. See Chapter 18.

CAT does not claim nonlinear singularity resolution. The singularity-resolution question is currently OPEN and, at the linearised level, VERIFIED negative. The nonlinear extension is the last roughly twenty percent chance of rescue, and the project's working estimate is pessimistic. See Chapter 18.

CAT does not claim peer-reviewed status of every result. The project has seven papers deposited on Authorea, with canonical DOIs, all currently in preprint form. "Verified" in this book always refers to the project's own eight-layer verification pipeline, not to peer review. See Chapter 20.

CAT does not claim to be the final theory. The book commits to one branch of one architecture. The project has not surveyed the full space of branches; it has not surveyed the full space of architectures. CAT is one consistent way to arrange the structure; the theorem-grade content is the arithmetic that follows from that arrangement. There may be others. See Chapter 23.

CAT does not promise that the canonical branch is right. The canonical branch is the branch the book commits to; it is the branch that all of Part II depends on. The blast-radius discipline of Chapter 10 makes the dependency explicit, and the five flagged measurements that could kill the canonical branch are listed there. If the canonical branch dies, Part II dies with it, and the work moves to the alternative branch. The strict gravitational sector survives in either case.

CAT does not claim that the Hasse-diagram observable is the unique discrete probe of curvature. The CJ observable of Chapter 5 is one Lorentz-invariant, path-based discrete probe of Weyl curvature, complementary to the Benincasa-Dowker discrete Ricci action. There may be others. The framework is built to admit them. The specific choice of CJ is the cleanest one currently known to the project, not a uniqueness theorem.

The discipline of this list is the same as the discipline of the verification pipeline: distinguish the strong claim from the weaker one, and never promote the second into the first.

CAT and its rivals, side by side

The following table is the one-page version of the question every reader of any theory-of-everything book asks: how does CAT compare to the others. The honest answer is that no two of the candidates on the market answer the same question, so a head-to-head table risks comparing things that are not, strictly, comparable. The table below is, with that caveat, a calibrated attempt.

Theory	Gravity sector (today)	SM & cosmology parameters	Falsifiable predictions	Empirical contact
CAT/SCT	[VERIFIED] one loop, parameter-free	$27 \to 1$ on canonical branch	ten-plus, within-decade reach	GWTC-3 (dispersion), Eot-Wash, Cassini
String theory	not closed	$\sim 10^{500}$ landscape	few; out-of-reach scales	none direct
Loop quantum gravity	not closed	sourced from outside	few; partial reach	none direct
Causal sets	action only	sourced from outside	few; within-decade	limited (EHT)

The honest reading of this table is that it is structured to make CAT look comparatively well off, which is the kind of bias the methodology chapters of Part V are explicitly trying to fight. So let me put the cards down on the table.

What CAT has that the others do not is a list of falsifiable numbers that are computed from the framework, not fitted from data. Thirty-plus of them, with status tags, with chapter-by-chapter derivations, and with experimental tests scheduled for the next decade. None of the rival theories on the table has produced a list of comparable specificity at the current state of the art.

What the rival theories have that CAT does not is forty years of work, thousands of citations, hundreds of practitioners, and the backing of a community that includes a substantial fraction of the brightest physicists alive. The single-author indie status of this project is a fact, not a strength; it is something the verification pipeline of Chapter 20 is specifically engineered to compensate for.

The sectors where CAT is strongest are also the ones where it is most exposed: a single-author project with a tight list of predictions has fewer places to hide than a community programme with a diffuse one. A measurement that lands the wrong way removes a chapter from this book. The rival theories are more diffuse in this sense; they have more dimensions of flexibility. Whether that flexibility is a virtue or a problem depends on which side of the demarcation line between physics and metaphysics one places the standards of the field.

The table is not the whole story. The whole story is in the chapters. The table is for the reader who wants the executive summary first, before deciding whether to read the chapters.

Frequently asked questions

The questions below are the ones I have been asked most often, either by friends, by colleagues, or by people who emailed me after reading the early preprints of the project. They are listed in roughly the order they tend to come up.

Is any of this peer-reviewed? Seven papers underlying this book are deposited on Authorea with canonical DOIs as preprints. Paper 7, on the Hasse-diagram approach to causal-set curvature, has its DOI listed in the Sources section. The eight-layer verification pipeline of Chapter 20 is a substitute, not a replacement, for peer review; it is what an indie theoretical physicist working without an institutional review pipeline can do, and the book is careful to label every result with the level of confidence it actually carries.

Are you a real physicist? I am an independent theoretical physicist with no formal institutional affiliation. I have a working knowledge of spectral-action calculations, heat-kernel methods, and the phenomenology of modified gravity, accumulated over the better part of a decade of self-study, with the assistance of every public textbook, lecture series, and arXiv preprint I could find. The work in this book has been cross-checked, where possible, against established literature, against three independent reasoning systems (the methodology of Chapter 22), and against eight verification layers. The reader is invited to evaluate the results, not the credentials.

What if you are wrong? The epilogue lists ten kill-switches, and the appendix *What would change my mind* on page 222 lists eleven softer Bayesian updates. Most of the kill-switches are scheduled for measurement within the next decade. If any of them flip, the relevant chapter of this book is going to have to be unwritten. The status tags VERIFIED, PROVEN, CONDITIONAL, OPEN, HEURISTIC, SPECULATIVE are designed precisely so that the reader can see which sectors are most exposed to which measurements. The strict gravitational sector is the most likely to survive; the canonical branch, conditional on it, is the most likely to fail.

Where is the code? The code base of the project is a Python package with about five thousand pytest tests, plus a Lean 4 formalisation harness, plus a tensor-algebra pipeline for heavy gamma-trace computations. The public-facing release is on the repository associated with the project's ORCID profile, reachable via the QR code below.

ORCID

Code that has not yet been released is held against future paper submissions, and will be released as the corresponding papers come out of preprint and into peer-reviewed venues. Reproducibility is a working principle, not a slogan; if a reader wants to verify a specific calculation in this book, the code is available on request.

How long did this take? The version of the project that exists today represents about fifteen months of full-time work, beginning in early 2025 and finalising in the spring of 2026, with the seven papers and the verification harness produced in that window. The reading-into-the-field period that preceded it was longer, roughly a decade, mostly part-time. The book itself was drafted in the four weeks before its deadline.

Is this falsifiable? Yes. The list of falsifiable predictions is the canonical-numbers appendix on page 234; the list of measurements that would falsify them is the *What would change my mind* appendix; the list of measurements that would falsify the entire theory is the kill-switch list in the epilogue. If, ten years from now, none of those measurements has flipped a single result, the case for the theory will have strengthened considerably. If one of them flips, the theory has to be rewritten or abandoned in the affected sector. Either outcome is consistent with the methodology.

Do you work alone? The mathematical work is single-author. The verification pipeline includes three independent reasoning collaborators (Chapter 22), whose role is closer to that of an adversarial referee than that of a co-author. The decision-making, the narrative, and the responsibility for every claim in this book are mine.

Is this a theory of everything? No. CAT is, on the canonical branch, a theory of the twenty-seven parameters of the Standard Model and cosmology that the canonical census theorem covers. It is not a theory of the absolute neutrino mass scale, of dark matter, of the inflationary mechanism, or of the global Lorentzian closure of quantum gravity. The book is careful about where the framework speaks and where it does not. See *What CAT does not claim* on page 225.

How do I cite the book? The book has its own ISBN, listed on the colophon page on the verso of the title page. The seven underlying papers have their own canonical DOIs, listed in *Sources for further reading* on page 256. For citing a specific result, cite the paper, not the book; the book is the pop-science companion, and the papers are the primary references.

Where do I report errors? Errors, large or small, can be reported to the author via the email QR code below.

Email author

The project maintains an errata page tied to each paper's DOI, and substantive corrections are reflected both in subsequent versions of the relevant paper and in subsequent editions of this book. For comments and suggestions on the book itself, the same address is fine.

What is next? The next year of the project's working life consists of attempting to close the open questions of Part IV: all-orders UV-finiteness, all-orders unitarity, the global Lorentzian closure, the nonlinear singularity question, and the three parallel routes to a derivation of v_{EW}. Some of these will close; some will not. The publication output, on the working estimate, is two to four papers per year over the next five years. The status tags will move; some chapters will be rewritten; the canonical-numbers appendix will be updated. The honest answer is that I do not know which of the open questions closes first, and the project's job is to find out.

Canonical numbers, at a glance

Every headline numerical prediction of CAT, in one place, with its status tag and a pointer back to the chapter where it is derived. For the reader who wants the version of the book that fits on one page, this is it.

A note on the tags. [VERIFIED] means the result has passed the project's eight verification layers (analytic, numerical to 100 digits, property fuzzing, literature comparison, dual derivation, triple computer-algebra cross-check, Lean formal proof, multi-backend agreement) plus three infrastructure layers (quick-sanity, consistency-DAG, provenance) — eleven independent checks in total. It does *not* mean the prediction has been measured in a laboratory. [PROVEN] means the result is a theorem with an explicit proof. [CONDITIONAL] holds given the canonical-branch axioms. [OPEN] means the question is part of the active research program and not yet closed.

Strict gravitational sector ($\mathcal{S}$**-layer**).

- [VERIFIED] $\alpha_C = \frac{13}{120}$: parameter-free Weyl coefficient at one loop (Chapter 1).
- [VERIFIED] $\alpha_R(\xi) = 2(\xi - \frac{1}{6})^2$: parameter-free Ricci coefficient at one loop (Chapter 2).
- [PROVEN] $\Pi_s(z, \xi) > 1$ for all $z > 0$, $\xi \neq 1/6$: no-scalaron theorem (Chapter 2).

- VERIFIED $m_2 = \sqrt{60/13}\,\Lambda \approx 2.148\,\Lambda$: spin-2 effective mass (Chapter 2).
- VERIFIED $\Lambda > 8.50\,\mathrm{meV}$ (95% CL): gravitational-wave dispersion bound from GWTC-3 (Chapter 4).
- VERIFIED $\Lambda > 3.53\,\mathrm{meV}$: Eot-Wash torsion-balance bound, strongest in PPN tier (Chapter 4).
- VERIFIED Standard Model is $4.7\times$ above critical fermion ratio and $12\times$ above critical vector ratio for no-scalaron robustness (Chapter 2).

Per-species heat-kernel coefficients.

- VERIFIED $\beta_W^{(0)} = +1/120$: per real scalar field (Chapter 3).
- VERIFIED $\beta_W^{(1/2)} = -1/20$: per Dirac fermion (with the famous fermion-loop minus sign) (Chapter 3).
- VERIFIED $\beta_W^{(1)} = +1/10$: per gauge boson (Chapter 3).
- VERIFIED $N_s = 4$, $N_D = 22.5$, $N_v = 12$: Standard Model field counts (Chapter 3).

Branch package ($\mathcal{B}$-layer, conditional on the canonical branch).

- CONDITIONAL $\mathcal{M} = \mathbb{C} \oplus \mathbb{C}$: Morgenstern algebra (Chapter 7).
- CONDITIONAL $\tau_\star = i\sqrt{2}$: modular point, a CM point of discriminant $D = -8$, corresponding to $\mathbb{Q}(\sqrt{-2})$ (Chapter 7).
- CONDITIONAL $\mathbb{Q}(\zeta_{48}) = \mathbb{Q}(\zeta_{48})$: cyclotomic carrier field for the modular relations (Chapter 7).
- CONDITIONAL $\lambda_\star = \varepsilon^{-2} = 3 - 2\sqrt{2}$, $\varepsilon^2 = 3 + 2\sqrt{2}$: small parameter of the flavour sector (Chapter 8).
- CONDITIONAL $\sin^2\theta_{13} = (3 - 2\sqrt{2})/8 \approx 0.0215$: PMNS reactor angle (Chapter 8).
- CONDITIONAL $\sin^2\theta_{12} + \sin^2\theta_{13} + \sin^2\theta_{23} = (23 - \sqrt{2})/24 \approx 0.8994$: PMNS sum rule (Chapter 8).
- CONDITIONAL $|J_{\mathrm{PMNS}}|^2 = (414 + 1223\sqrt{2})/2{,}359{,}296 \approx 9.09 \times 10^{-4}$: Jarlskog modulus squared (Chapter 8).

- CONDITIONAL $\delta_{\rm CKM} = 3\pi/8 = 67.5°$: quark CP-violating phase (Chapter 8).
- CONDITIONAL $\sin^2\theta_{23} = (6-\sqrt{2})/8 \approx 0.5732 > 1/2$: atmospheric mixing forced into upper octant in the canonical branch (Chapter 10).
- CONDITIONAL Twenty-seven-to-one parameter collapse: every Standard-Model and cosmological parameter except the electroweak scale $v_{\rm EW}$ is fixed by the canonical branch (Chapter 9).

Cosmological sector.

- CONDITIONAL $\Omega_\Lambda = 168/(25\pi^2) \approx 0.6809$: dark-energy fraction; $168 = 24 \cdot 7 = |\mathrm{PSL}(2,7)|$ (Chapter 11).
- VERIFIED $S = A/4G + 13/(120\pi) + (37/24)\ln(A/\ell_P^2)$: black-hole entropy with two CAT corrections, the constant $13/(120\pi)$ tracing back to $\alpha_C = 13/120$ (Chapter 14).
- VERIFIED $\delta H^2/H^2 \approx 1.3 \times 10^{-64}$: modified-cosmology MT-2 result, parametrically suppressed (Chapter 15).
- OPEN Dark-matter prediction: the canonical branch does not provide a specific dark-matter candidate (Chapter 12).
- OPEN Inflation mechanism: dilaton route excluded, two-scale framework mandatory but not yet a working mechanism (Chapter 13).

Open-program markers ($\mathcal{P}$-layer).

- OPEN All-orders UV-finiteness: chiral-Q route conditional on Gap G1 closure (Chapter 17).
- OPEN All-orders unitarity: fakeon prescription primary, conditional on global-positivity GP (Chapter 17).
- OPEN Nonlinear singularity resolution: NEGATIVE at linearised level, $\sim 20\%$ chance non-linear extension rescues it (Chapter 18).
- OPEN Global Lorentzian closure: open program (Chapter 18).
- OPEN $v_{\rm EW}$ derivation: three parallel routes (A: Nonlocal Transmutation, C: Modular L-values, E: Sigma Mechanism Derived); combined success probability $\sim 66\%$ by working estimate (Chapter 19).

Causal-set toolkit.

- VERIFIED Hasse diagram is the carrier of geometric information about a sprinkled causal set (Paper 7, Chapter 5).
- VERIFIED CJ observable: Lorentz-invariant, path-based, sensitive to Weyl curvature, complementary to the Benincasa-Dowker action (Chapter 5).
- VERIFIED Boost test for CJ: $E^2 \propto$ CJ pattern, B-squared-blind to better than 1%, flat to sub-percent precision (Chapter 5).

Operational rules from the methodology.

- Sectoral derivative principle: $\partial \alpha_C / \partial \theta = 0$ for any branch parameter θ (Chapter 10).
- Eight verification layers plus three infrastructure layers for every VERIFIED or PROVEN result (Chapter 20).
- Catalogue of fifteen failure modes from FND-1 phenomenology (Chapter 21).
- Three-mind protocol: six mandatory gates (seeding, code review, de-duplication, demolition, analytical confirmation, tribunal), as described in Chapter 22.

Notation

The symbols used in this book, in roughly the order in which they first appear, with their meaning in one short phrase. For longer definitions, see the glossary on page 242; for the headline numerical values, see the canonical-numbers appendix on page 234.

Strict gravitational sector.

α_C. Weyl coefficient at one loop, equal to $13/120$. The headline number of the gravity sector.

$\alpha_R(\xi)$. Ricci coefficient at one loop, equal to $2(\xi - 1/6)^2$. Vanishes at conformal coupling.

Λ. CAT energy scale at which the form factor turns on. Bounded from below by experiment to be greater than $8.50\,\text{meV}$.

ξ. Higgs non-minimal coupling to gravity. Free in the $\mathcal{S}$-layer; fixed by the canonical branch.

$\varphi(z)$. Master function controlling the form factor. Equal to $e^{-z/4}\sqrt{\pi/z}\,\text{erfi}(\sqrt{z}/2)$.

$\Pi_s(z,\xi)$. Scalar-block propagator denominator. Strictly greater than one for $z > 0, \xi \neq 1/6$.

$\Pi_{\text{TT}}(z)$. Transverse-traceless-block propagator denominator. Has a single positive zero at the spin-2 mass.

m_2. Spin-2 effective mass, equal to $\sqrt{60/13}\,\Lambda \approx 2.148\,\Lambda$.

$F_1(z), F_2(z, \xi)$. Total form factors, sums over Standard-Model species.

$\beta_W^{(s)}, \beta_R^{(s)}$. Per-spin coefficients of the Weyl and Ricci pieces, $s = 0, 1/2, 1$.

N_s, N_D, N_v. Standard-Model field counts: real scalars (4), Dirac equivalents (22.5), gauge bosons (12).

Branch package.

$\mathcal{M}$. Morgenstern algebra, $\mathbb{C} \oplus \mathbb{C}$.

$\tau_\star$. Modular point of the canonical branch. Equal to $i\sqrt{2}$. A CM point of discriminant $D = -8$.

$\mathbb{Q}(\zeta_{48})$. Cyclotomic field $\mathbb{Q}(\zeta_{48})$.

ε. Small parameter of the flavour sector, $\varepsilon^2 = 3 + 2\sqrt{2}$.

$\lambda_\star$. Reciprocal small parameter, $\lambda_\star = \varepsilon^{-2} = 3 - 2\sqrt{2}$.

$\sin^2\theta_{12,13,23}$. PMNS mixing angles squared, neutrino sector.

$\sin^2\theta^q_{12,13,23}$. CKM mixing angles squared, quark sector.

J_{PMNS}. Jarlskog invariant of the PMNS matrix.

δ_{CKM}. CP-violating phase of the CKM matrix, equal to $3\pi/8 = 67.5°$ in the canonical branch.

Cosmological sector.

Ω_Λ. Dark-energy fraction, equal to $168/(25\pi^2) \approx 0.6809$ in the canonical branch.

H. Hubble expansion rate.

$\delta H^2/H^2$. Modified-cosmology fractional correction to the Hubble rate squared. About 1.3×10^{-64} at present-day densities.

S. Black-hole entropy. CAT prediction: $S = A/4G + 13/(120\pi) + (37/24)\ln(A/\ell_P^2)$.

A. Black-hole horizon area.

ℓ_P. Planck length.

M_P. Planck mass.

$v_{\rm EW}$. Higgs vacuum expectation value, about 246 GeV. Sets the scale of the weak nuclear force.

Causal-set toolkit.

CJ. The CAT path-based observable on the Hasse diagram of a causal set. Detects Weyl curvature.

E^2, B^2. Components of the Weyl curvature projected on a foliation: the electric (gravitoelectric) and magnetic (gravitomagnetic) parts, respectively.

Layer labels.

$\mathcal{S}$-**layer.** Strict gravitational sector. Theorem-grade. Part I.

$\mathcal{B}$-**layer.** Branch package. Theorem-grade conditional on the canonical branch. Part II.

$\mathcal{P}$-**layer.** Open program. Part IV.

$\partial\alpha_C/\partial\theta$. Sectoral derivative. Equal to zero for any branch parameter θ.

Verification status tags.

[proven]. Theorem-grade. Eight-layer verification including formal-proof layers. The strongest tag.

[verified]. Numerical and analytic agreement across all eight verification layers, but without a formal proof.

[conditional]. Theorem-grade conditional on a stated assumption (typically the canonical branch).

[open]. Open question. Not closed by the present version of the theory.

heuristic. Heuristic argument or working estimate. Not a verified result.

speculative. Speculative. Mentioned but not endorsed.

Glossary

The special vocabulary used in this book, in alphabetical order and in plain English. Each entry points to the chapter where the term is first defined, for the reader who wants the longer explanation.

α_C **(Chapter 1)**. The parameter-free coefficient that controls the leading quantum correction to the Weyl part of the gravitational form factor in CAT. Equal to $13/120$. The Weyl part is what gravitational waves are made of.

$\alpha_R(\xi)$ **(Chapter 2)**. The parameter-free coefficient that controls the leading quantum correction to the Ricci part of the gravitational form factor. Equal to $2(\xi - 1/6)^2$, a perfect square that vanishes at the conformal-coupling value $\xi = 1/6$.

AQL axiom (Chapter 8). A constraint that the canonical CAT branch imposes on flavour observables: each observable mixing must be expressible as an algebraic number in the field $\mathbb{Q}(\sqrt{2})$. Forces the PMNS and CKM matrices to clean closed-form expressions.

Benincasa-Dowker action (Chapter 5). The standard discrete observable in causal set theory for extracting the Ricci scalar curvature from the combinatorial structure of a sprinkled causal set. Workhorse of the program since 2010. Complementary to the CJ observable.

Blast radius (Chapter 10). A measure of how much of CAT dies if a particular experimental result comes in. Three sizes: small (single prediction), medium (canonical branch), large (strict gravitational sector).

Branch (Chapter 7). A discrete binary choice that distinguishes one half of the Morgenstern algebra from the other. The canonical branch is the one this book commits to; the alternative branch produces incompatible predictions for the flavour and cosmological parameters.

Causal set (Chapter 5). A discrete combinatorial object consisting of a finite collection of events with causal relations among them. The central object of causal set theory, a candidate framework for fundamental spacetime.

Census (Chapter 9). The canonical census theorem (v3.0). The statement that twenty-seven free parameters of the Standard Model and cosmology collapse to one (the electroweak scale v_{EW}) within the canonical branch of CAT.

CJ observable (Chapter 5). A Lorentz-invariant, path-based statistic on the Hasse diagram of a causal set that detects Weyl curvature. The subject of Paper 7. Complementary to the Benincasa-Dowker action.

CKM matrix (Chapter 8). The Cabibbo-Kobayashi-Maskawa mixing matrix for quarks. Has three angles and one CP-violating phase. In the canonical branch, all four parameters fall out of the modular structure.

CM point (Chapter 7). A complex-multiplication point. A particular point on the modular fundamental domain at which modular functions evaluate to algebraic numbers. The canonical branch's modular point $\tau_\star = i\sqrt{2}$ is a CM point of discriminant $D = -8$.

Conformal coupling (Chapter 2). The specific value $\xi = 1/6$ of the Higgs non-minimal coupling to gravity at which the scalar sector decouples from gravity. At this value, α_R vanishes exactly.

Cosmological constant problem. The standard puzzle of why Ω_Λ is the value it is, roughly 0.685, rather than 10^{60} or 10^{120} orders of magnitude larger as naive estimates suggest. The canonical branch sidesteps the puzzle by computing Ω_Λ from modular arithmetic; see Chapter 11.

Dirac equivalents (Chapter 3). A normalisation convention for counting Standard-Model fermions. Two Weyl fields equal one Dirac field. The Standard Model has 22.5 Dirac equivalents, fifteen per generation times three generations divided by two.

ε **(Chapter 8).** The small parameter of the canonical-branch flavour sector, defined by $\varepsilon^2 = 3 + 2\sqrt{2}$. Appears as the expansion parameter in the closed-form expressions for the PMNS and CKM mixings.

Fakeon prescription (Chapter 17). The Anselmi prescription for handling would-be ghost modes in higher-derivative theories of gravity. The primary resolution mechanism for the unitarity question in CAT, conditional on a property the project labels GP.

Form factor (Chapter 1). The function that describes how a quantum field is modified at short distances by the presence of other quantum matter. In CAT, the gravitational form factor is computed at one loop and contains the parameter-free coefficients α_C and α_R.

FND-1 (Chapter 21). Fundamental Investigation 1, the project's first phenomenology campaign, run in late 2025 and early 2026. Produced eighteen overclaim corrections that the verification pipeline caught.

Hasse diagram (Chapter 5). The standard graph representation of a finite partial order: one node per event, one edge per direct causal relation, with indirect relations not drawn. The canonical carrier of geometric information about a causal set, established in Paper 7.

Hierarchy problem (Chapter 19). The standard puzzle of why v_{EW}/M_P is roughly 10^{-17}, seventeen orders of magnitude. Forty years of candidate mechanisms have run aground on it. CAT's three parallel routes attempt to derive the ratio from modular structure.

Jarlskog invariant (Chapter 8). A single rephasing-invariant quantity that captures all of the CP-violating physics of a flavour mixing matrix. In the canonical branch, $|J_{\mathrm{PMNS}}|^2 = (414 + 1223\sqrt{2})/2{,}359{,}296$.

Kugo–Ojima quartet (Chapter 17). A standard mechanism in gauge theory for packaging unphysical modes into a balanced quartet that decouples from physical observables. Rejected as a resolution for the unitarity question in CAT, on a structural ground specific to the higher-derivative setting.

Λ **(Chapter 1).** The energy scale at which the CAT gravitational form factor turns on. The only free parameter in the strict gravitational sector. Bounded from below by experiment to be greater than $8.50\,\mathrm{meV}$; the next decade of measurements will either tighten the bound or measure Λ outright.

Master function φ **(Chapter 1).** The single curve that controls the gravitational form factor. Computed once from the Standard-Model field content; not a free parameter.

Modular point (Chapter 7). The complex number $\tau_\star$ at which the canonical branch's modular relations are evaluated. In the canonical branch, $\tau_\star = i\sqrt{2}$.

Morgenstern algebra (Chapter 7). The algebra $\mathcal{M} = \mathbb{C} \oplus \mathbb{C}$ that hosts the canonical CAT branch. The simplest algebra that admits a binary discrete choice without continuous deformation.

Open-program items. The five major open questions in the $\mathcal{P}$-layer: global Lorentzian closure, all-orders unitarity (the fakeon-versus-Kugo-Ojima question), all-orders UV-finiteness, graviton scattering at one loop, and nonlinear singularity resolution. See Chapter 17 and Chapter 18.

Naked notes. The italic, dated, first-person interludes inside almost every chapter of this book. A methodological commitment, not a stylistic flourish; see Chapter 23.

No-scalaron theorem (Chapter 2). The result that, in the canonical CAT, no real scalar particle propagates in the gravitational sector. Verified at the linearised level for the Standard-Model matter content (April 2026), through the eight-layer pipeline of Chapter 20.

Ω_Λ **(Chapter 11).** The fraction of the present universe's energy budget that consists of dark energy. The canonical-branch prediction is $168/(25\pi^2) \approx 0.6809$; the Planck 2018 measurement is 0.6847 ± 0.0073.

PMNS matrix (Chapter 8). The Pontecorvo-Maki-Nakagawa-Sakata mixing matrix for neutrinos. In the canonical branch, all of its angles and the CP phase are algebraic numbers in $\mathbb{Q}(\sqrt{2})$.

Π_s **(Chapter 2).** The denominator of the scalar-block propagator in CAT. Strictly greater than one for all positive momenta and all values of the Higgs coupling, by the no-scalaron theorem.

Π_{TT} **(Chapter 2).** The denominator of the transverse-traceless-block propagator in CAT. Has a single positive zero, located at the spin-2 effective mass $m_2 = \sqrt{60/13}\,\Lambda$.

RAZUMIST (Chapter 22). The project's adversarial multi-model methodology. Six mandatory gates through which every non-trivial result passes before being committed to the canonical-results database.

$\mathcal{S}/\mathcal{B}/\mathcal{P}$**-layer (Prologue).** The three layers of CAT. $\mathcal{S}$: gravitational sector, theorem-grade, Part I. $\mathcal{B}$: branch package, theorem-grade conditional on the canonical branch, Part II. $\mathcal{P}$: open program, Part IV.

Sectoral derivative (Chapter 10). The principle $\partial \alpha_C / \partial \theta = 0$ for any branch parameter θ. Says that the strict gravitational sector is decoupled from the branch package, so any branch death leaves the gravity story intact.

Sprinkling (Chapter 5). The procedure for generating a causal set from a Lorentzian manifold by Poisson-distributing events at a chosen density and inheriting the causal relations.

$\tau_\star$ **(Chapter 7).** The modular point of the canonical branch. Equal to $i\sqrt{2}$. A CM point of discriminant $D = -8$.

v_{EW} **(Chapter 19).** The Higgs vacuum expectation value, about 246 gigaelectronvolts. Sets the scale of the weak nuclear force.

Verification pipeline (Chapter 20). The eight-layer verification process plus three infrastructure layers that every [VERIFIED] or [PROVEN] result in the canonical-results database has passed.

Weyl curvature (Chapter 5). The trace-free part of the full Riemann curvature tensor. Describes how gravitational disturbances propagate through empty regions of spacetime. What gravitational waves are made of.

ξ **(Chapter 2).** The Higgs non-minimal coupling to gravity. Free continuous parameter in the strict $\mathcal{S}$-layer; fixed by the canonical branch in Part II via the census theorem.

$\mathbb{Q}(\zeta_{48})$ **(Chapter 7).** The cyclotomic field $\mathbb{Q}(\zeta_{48})$ generated by the 48th roots of unity. The smallest field of complex numbers in which the canonical branch's modular relations can be expressed.

Notes

The following notes collect the asides, technical caveats, and small clarifications that did not fit into the main text of each chapter without breaking the flow. They are organised chapter by chapter, with brief signposts. None of the notes is required for the main argument; they are for the reader who wants the slightly longer answer.

* * *

Notes to Chapter 1 (The shape of gravity, up close)

The form factor and the Wilsonian effective action. What I called the "form factor" is, in the working language of quantum field theory, a piece of the Wilsonian one-loop effective action: the part that survives when you integrate out the matter loops of the Standard Model and ask what is left for the metric. The piece is universal, parameter-free, and computable, which is why the chapter spent the time it did on the calculation rather than on a vague gesture toward it.

Why α_C is a coefficient and not a free parameter. The standard objection a reader of effective-field-theory textbooks might raise is that the coefficients of higher-derivative terms in gravity are free parameters, fitted from experiment. That is true for the bare action. The CAT setup computes the α_C that appears in the one-loop *quantum* correction, which is not the bare-action coefficient but the coefficient of the running piece. The bare-action piece is renormalised away by the counter-term; what survives is the universal one-loop piece, and that is what the chapter computes.

* * *

Notes to Chapter 2 (No-scalaron, and the absence of a fifth force)

Why "no-scalaron" is not the same as "no scalar field". The no-scalaron theorem says no *propagating* real scalar particle appears in the gravitational sector. The Higgs is, of course, a scalar field; it is non-minimally coupled to gravity through its ξ coupling, and it does propagate, but it is a matter field, not a gravitational degree of freedom. The theorem is about the gravitational sector, where a scalar mode would mean a scalar partner of the graviton, and the chapter is careful about the distinction.

Why $\xi = 1/6$ is not arbitrary. The conformal-coupling value $\xi = 1/6$ in four dimensions is the value at which a free scalar field's energy-momentum tensor becomes traceless (modulo conformal anomalies). It is the unique value at which the kinetic action of the scalar is conformally invariant, and it is the value at which the scalar decouples cleanly from the gravitational sector. The canonical branch fixes it; the strict gravitational sector treats it as free.

* * *

Notes to Chapter 3 (Where 13/120 comes from)

The fermion-loop minus sign, in plain language. The famous minus sign in the per-Dirac fermion contribution comes from the fact that fermions, in the path-integral formulation, are integrated as anticommuting (Grassmann) variables. The Gaussian integral over Grassmann variables produces a determinant in the numerator of the partition function, where the bosonic Gaussian integral produces a determinant in the denominator. The log-determinant of the fermion side picks up an opposite sign, and that is where the $-1/20$ comes from.

Why $N_D = 22.5$ rather than 45. A reader who counts Standard-Model fermions naively gets fifteen Weyl spinors per generation, times three generations, equals forty-five. The bookkeeping convention used here counts *Dirac equivalents*: two Weyl spinors equal one Dirac field, so 45 Weyl spinors equal 22.5 Dirac equivalents. The half-integer number reflects the fact that the Standard Model has chiral matter, and the sector is not Dirac-symmetric.

* * *

Notes to Chapter 4 (The bound on Λ)

Why GW170817 is not the strongest test. The 2017 binary-neutron-star event GW170817 measured the speed of gravitational waves to extreme precision, and famously killed several modified-gravity dark-energy models. CAT survives that test cleanly, because the transverse-traceless block of CAT is massless at zero momentum. The strongest CAT-specific bound comes not from the speed of the waves but from their dispersion: how the group velocity varies with frequency. That is a GWTC-3 catalogue analysis, not a single-event measurement, and it is what gave the $\Lambda > 8.50\,\text{meV}$ bound.

* * *

Notes to Chapter 5 (Curvature from a discrete universe)

Why the boost test is so important. A discrete observable that depended on a frame of reference would be useless in a Lorentz-invariant theory: the answer would change when you boosted to a different frame. The Benincasa-Dowker action passes the boost test by construction; the CJ observable passes it as an empirical result. Paper 7 reports the explicit numerical check: $E^2 \propto \text{CJ}$, with the B^2 contribution suppressed below the 1% level over the boost range tested. Without that result, the CJ observable would not be a candidate for a Lorentz-invariant discrete theory of gravity.

* * *

Notes to Chapter 6 (A handedness in the heat trace)

Why the chiral-Q quantization is finer than D^2. The standard D^2 quantization computes the heat trace of a single second-order operator and sums over chirality implicitly. The chiral-Q quantization splits the operator into two chirality-resolved blocks and computes their heat traces separately. The two prescriptions agree on the chirality-summed coefficients, but the chiral-Q quantization preserves information that the D^2 quantization throws out. Specifically, it preserves the relations between the two blocks at every loop order, which is what makes the loop-independence property emerge.

* * *

Notes to Chapter 7 (The branch, in plain language)

Why $\mathbb{C}\oplus\mathbb{C}$ rather than just $\mathbb{C}$. The Morgenstern algebra has to admit a binary discrete choice, which a single copy of $\mathbb{C}$ does not. The minimal algebra that does is the direct sum of two copies, with the discrete choice corresponding to which factor a state lives in. Larger algebras would also work, but the project's economy principle selects the smallest one consistent with the structure of the canonical branch.

* * *

Notes to Chapter 8 (Three matrices you can compute)

Why the AQL axiom is not arbitrary. The AQL axiom says that observable mixings live in $\mathbb{Q}(\sqrt{2})$. In a generic flavour theory there is no reason for the mixings to live in a number field at all; they could be transcendental, or rational with no algebraic structure. The AQL axiom is the constraint that the canonical-branch theory imposes, and it is the constraint that closes the flavour sector into a finite set of algebraic numbers. Without it, the twenty-seven-to-one census fails.

* * *

Notes to Chapter 9 (The census, formally)

Version 29. The canonical census theorem has gone through twenty-eight earlier versions, each of which was either falsified by a later calculation (the most common outcome) or replaced by a stronger formulation. Version 29 is the one that survives the current verification pipeline. There is no guarantee that version 30 will not replace it.

* * *

Notes to Chapter 11 (Dark energy, a number that landed)

Why $168 = 24 \cdot 7$. The numerator 168 in $\Omega_\Lambda = 168/(25\pi^2)$ is the order of the simple group $\mathrm{PSL}(2,7)$, which is the smallest non-trivial finite simple group of Lie type after the alternating group on five letters. The relevance of this group to the canonical branch is one of the open structural mysteries of the project, in the sense that the relationship is verified numerically but not yet derived from a clean first principle.

* * *

Notes to Chapter 14 (Black-hole entropy)

Why the logarithmic correction is $37/24$. Both correction terms in the black-hole entropy formula (the constant $13/(120\pi)$ and the logarithmic $(37/24)\ln(A/\ell_P^2)$) are computed for the canonical Standard-Model field content, the same content that gave $\alpha_C = 13/120$ in Part I. The number $37/24$ comes from the per-species heat-kernel coefficients of the matter sector on a Schwarzschild background, summed with the same field counts ($N_s = 4$, $N_D = N_f/2 = 22.5$, $N_v = 12$) used throughout. The SM $+ 3\nu_R$ extension shifts the constant correction by a calculable amount, separately from the logarithmic coefficient; both shifts are recorded in the project's verification database.

* * *

Notes to Chapter 16 (The universe, expanding)

Why CAT cosmology cannot have observable deviations at present-day densities. The form factor $\varphi(\Box/\Lambda^2)$ acts on the metric. On the FLRW background, $\Box$ becomes H^2, and the present-day Hubble rate squared is roughly $10^{-66}\,\mathrm{eV}^2$. With Λ bounded above 8.50 meV, the dimensionless argument H^2/Λ^2 is at most 10^{-64}. The form factor at that argument is essentially $\varphi(0) = 1$, with corrections of the same order. The universe is too cold and too large for the modification to do anything observable. This is, on reflection, exactly what a sensible modified-gravity theory should say.

* * *

Notes to Chapter 17 (UV-finiteness and unitarity)

The Anselmi fakeon prescription, in plain language. A fakeon is a would-be ghost mode of the theory whose propagator is treated by a specific analytic continuation, instead of being either thrown away (which would break unitarity) or kept (which would break stability). The continuation removes the negative-norm component of the would-be ghost while preserving the unitarity of the physical S-matrix. CAT inherits the prescription from the fakeon-quantisation programme of Anselmi and others.

* * *

Notes to Chapter 19 (v_{EW} parallel routes)

Why three routes rather than one. The hierarchy problem has resisted forty years of single-strategy attacks, and the project's working estimate is that a single CAT route has only a small chance of closing it. Three parallel routes, each with an independent $\sim 3\%$–4% probability, give a combined success probability of about 66% by elementary union-bound estimation, conditional on the routes being independent of each other. Whether they are independent is itself an open question.

* * *

Notes to Chapter 22 (Three minds, one answer)

Why three rather than two. A single adversarial collaborator is, in practice, not adversarial enough; it tends to defer. Two collaborators with different biases (analytical and synthetic) are stronger, but they produce deadlocks. Three collaborators with three different biases (analytical, synthetic, and arbitrative) avoid the deadlock and produce, in the project's experience, a substantially harder review than any single collaborator alone. The methodology is described in detail in the chapter; here I only note that the choice of three rather than two or four is empirical, not principled.

Further reading

For the reader who wants to go deeper, into the canonical literature of the surrounding fields, the list below is a curated selection. None of these books or papers is required for following the present work. Together they cover the surrounding territory.

General relativity, in textbook form. Wald, *General Relativity* (1984). The graduate-level standard, short and exact. Misner, Thorne, and Wheeler, *Gravitation* (1973). The encyclopaedic alternative, a thousand pages of beautiful diagrams. Carroll, *Spacetime and Geometry* (2004). The most accessible of the three for a determined reader.

Quantum field theory in curved spacetime. Birrell and Davies, *Quantum Fields in Curved Space* (1982). The original textbook. Parker and Toms, *Quantum Field Theory in Curved Spacetime* (2009). The modern update, with chapters on the effective action and the heat kernel.

Heat kernel and form factors. Vassilevich, *Heat kernel expansion: user's manual, Physics Reports* 388 (2003) 279. The standard reference for Seeley-DeWitt coefficients, including the curved-space conventions used in this book. Avramidi, *Heat Kernel and Quantum Gravity* (2000). The longer, more mathematical alternative. Codello and Percacci, *Fixed points of higher-derivative gravity, Physical Review Letters* 97 (2006) 221301. The foundational paper for the higher-derivative side that CAT generalises.

Spectral action and noncommutative geometry. Connes and Marcolli, *Noncommutative Geometry, Quantum Fields and Motives* (2008). The full noncommutative-geometry programme. Chamseddine and Connes, *The*

Spectral Action Principle, Communications in Mathematical Physics 186 (1997) 731. The foundational paper of the spectral-action approach that CAT extends.

Causal sets. Sorkin, *Causal sets: discrete gravity, arXiv:gr-qc/0309009* (2003). The foundational programme. Surya, *The causal set approach to quantum gravity, Living Reviews in Relativity* 22 (2019) 5. The modern review. Dowker, the body of work on the Benincasa-Dowker action and on Lorentz invariance of causal sets, against which the CJ observable of Chapter 5 should be read.

String theory. Polchinski, *String Theory*, two volumes (1998). The graduate-level standard. Becker, Becker, and Schwarz, *String Theory and M-Theory* (2007). The textbook with M-theory and braneworlds. Greene, *The Elegant Universe* (1999). The accessible non-technical introduction.

Loop quantum gravity. Rovelli, *Quantum Gravity* (2004). The textbook from one of the founders. Thiemann, *Modern Canonical Quantum General Relativity* (2007). The encyclopaedic alternative.

Asymptotic safety. Reuter and Saueressig, *Quantum Gravity and the Functional Renormalisation Group* (2019). The textbook for the asymptotic-safety programme. The negative result of Route D in Chapter 19 is a small contribution to the dialogue between asymptotic safety and CAT.

Modular forms and number theory. Diamond and Shurman, *A First Course in Modular Forms* (2005). The graduate textbook. Koblitz, *Introduction to Elliptic Curves and Modular Forms* (1993). The shorter alternative. Background for the modular-arithmetic content of Part II.

General-audience books on the same family of problems. Smolin, *The Trouble with Physics* (2006). The classic critique of the string programme, much of which CAT shares. Penrose, *The Road to Reality* (2004). The encyclopaedic field guide for the determined non-specialist, with derivations included. Carroll, *Something Deeply Hidden* (2019). The accessible introduction to the foundations of quantum mechanics, which is in the background of any quantum-gravity discussion. Hossenfelder, *Lost in Math* (2018). The critique of theoretical physics' aesthetic biases, which the methodology chapters of Part V take seriously.

The seven papers underlying this book. For the primary references behind the chapters of this book, see *Sources for further reading* on page 256, where each chapter is mapped to its underlying paper with the canonical DOI.

Sources

This book is a pop-science companion to a sequence of formal publications produced by the project between mid-2025 and early 2026. The mapping below makes the correspondence explicit. Each chapter cites its primary published source with DOI; readers who want the full technical detail behind any pop-science argument in this book should follow the citation to the corresponding paper.

The chapters are not derivative of the papers in the sense that they reproduce them. They are derivative in the sense that they offer a pop-science account of what the papers establish, in the same voice and following the same architectural choices, with naked-notes commentary on what the work was actually like.

Part I: SCT, Gravity ($\mathcal{S}$-layer)

- **Chapter 1** (The shape of gravity, up close) and **Chapter 3** (Where 13/120 comes from):
  Paper 1, *Nonlocal one-loop form factors of the spectral action with Standard Model content*, SSRN preprint, April 2026.

Paper 1

- **Chapter 2** (No-scalaron):
Companion paper, *No-scalaron theorem and the absence of a fifth force in the linearised Standard-Model gravitational sector*, in preparation; canonical-results verification in the project's working notes (NT-1b Phase 3, March 2026).

- **Chapter 4** (The bound on Λ from gravitational waves):
Paper 2, *Solar system and laboratory tests of the Spectral Causal Theory*.

Paper 2

- **Chapter 5** (Curvature from a discrete universe):
Paper 7, *Weyl curvature from the Hasse diagram: a parameter-free bridge formula for causal sets*, Alfyorov & Shnyukov, SSRN preprint, March 2026 (DOI: `10.2139/ssrn.6506701`).

Paper 7

Part II: Branch Package ($\mathcal{B}$-layer)

- **Chapter 7** (The branch, in plain language) and **Chapter 8** (Three matrices you can compute):
Paper 3, *Chirality of the Seeley-DeWitt expansion in Spectral Causal Theory*; plus the canonical-branch supporting derivations on the project's working drive.

Paper 3

- **Chapter 9** (The census, formally) and **Chapter 10** (Blast radius): Paper 5, *The predictive content of Spectral Causal Theory.*

Paper 5

Part III: Cosmology and the Universe

- **Chapter 11** (Dark energy: a number that landed) and **Chapter 15** (One prediction the theory got wrong):
  Paper 4, *Nonlinear field equations and FLRW cosmology in CAT*; plus Paper 8, *Non-perturbative spectral gravity measure in the Hilbert-Schmidt Gaussian completion: pro-torsor structure and the obstruction to canonical expectations*, Alfyorov & Shnyukov, SSRN preprint, April 2026 — the formal write-up of the alternative measure-based route that closes negative in the diagonal Hilbert-Schmidt sector.

Paper 4

Paper 8

- **Chapter 12** (Dark matter: a hole in the prediction list): Companion analysis, `dm-extension-chatgpt.md` (April 2026), and the SM + $3\nu_R$ extension document.
- **Chapter 13** (Inflation: not yet a mechanism): Companion analyses INF-2 and the Phase 1–13 early-universe programme (April 2026); Paper 6, *Auxiliary boundary data for inflationary CAT cosmology*, in preparation.
- **Chapter 14** (Black-hole entropy): MT-1 result document (April 2026), companion paper in preparation.

Part IV: Open Frontiers ($\mathcal{P}$-layer)

- **Chapter 17** (UV finiteness + unitarity): the all-orders UV-finiteness and fakeon-versus-Kugo-Ojima working analyses; project notes `op07-fakeon-analysis.md` and `mr5-finiteness.md`.
- **Chapter 18** (Curved spacetime: singularities + Lorentzian closure): the singularity-resolution and Lorentzian-closure working analyses; project notes `mr9-singularity.md` and `lt3e-tov.md`.
- **Chapter 19** (Where the weak force gets its scale): `theory/derivations/v_EW_parallel_routes.md`, plus `as-route-d.md` (closure of Asymptotic Safety route, April 2026).

Part V: Method

- **Chapter 20** (How do you know when you're right): Project's eight-layer verification pipeline documentation and accompanying Python and Lean code base.
- **Chapter 21** (Fifteen ways to fool yourself): Project's experiment-protocol notes and the FND-1 phenomenology campaign records of late 2025 and early 2026.
- **Chapter 22** (Three minds, one answer): Project's RAZUMIST protocol document (version 3.0.3) and the worked-example run logs.

- **Chapter 23** (Honest theory in the LLM era):
  Project's working naked-notes archive; the present book is itself the canonical reference.

Falsification atlas

- **Epilogue** (Where the theory could die):
  Built on the blast-radius semantic of Chapter 10, with each kill-switch traced back to its canonical-results entry in the project's working database.

Author scientific identifier
The full list of the project's publications is maintained on the author's ORCID profile, reachable via the QR code below, with cross-links to the canonical published versions.

ORCID profile

The project's working notes, including the canonical-results database and the full verification logs, are not publicly available but can be made available for spot-checking under appropriate confidentiality arrangements; the author's contact details are on the colophon page at the front of the book.

About the author

David Alfyorov is an independent theoretical physicist working in causal-architecture theory and one-loop quantum gravity. He spent the better part of a decade reading himself into the field without formal institutional affiliation, and since 2025 he has been producing the work described in this book from a kitchen table with a laptop and an unreasonable amount of cold tea.

The project that produced *Schrödinger's CAT* is his first sustained body of original theoretical work and represents, as the methodology chapters of Part V try to make explicit, an experiment in independent research carried out at the scale of a solo theoretical physicist with the assistance of contemporary computational tools.

The formal results that this book draws on were developed in ongoing collaboration with Igor Shnyukov, whose contribution to the canonical-results corpus is gratefully acknowledged. The prose, the framing, and any errors of judgement in this book are the author's alone.

His scientific publications are catalogued at his ORCID profile, reachable via the QR code below, where the formal papers underlying this book are listed with their canonical DOIs.

ORCID profile

The author may be reached via the email QR code below; correspondence concerning the canonical-results database, the verification logs, or any specific naked-notes block in this book is welcome.

Email author

The next decade of the project's working life is the one in which the predictions of *Schrödinger's CAT* will either survive their experimental tests or be replaced by something better. He intends to be there for the answer either way.

www.ingramcontent.com/pod-product-compliance
Lightning Source LLC
LaVergne TN
LVHW091115080826
845145LV00008B/1922

9781806451623